NEW WORLD COMING

NEW WORLD COMING

FRONTLINE VOICES ON PANDEMICS, UPRISINGS, & CLIMATE CRISIS

EDITED BY
ALASTAIR LEE BITSÓÍ
& BROOKE LARSEN

TORREY HOUSE PRESS
Salt Lake City • Torrey

First Torrey House Press Edition, November 2021

Published by Torrey House Press
Salt Lake City, Utah
www.torreyhouse.org

International Standard Book Number: 978-1-948814-53-9
E-Book ISBN: 978-1-948814-54-6
Library of Congress Control Number: 2021930607

Cover art by Mariella Mendoza
Cover design by Kathleen Metcalf
Interior design by Rachel Buck-Cockayne
Distributed to the trade by Consortium Book Sales and Distribution

Torrey House Press offices in Salt Lake City sit on the homelands of Ute, Goshute, Shoshone, and Paiute nations. Offices in Torrey are on the homelands of Southern Paiute, Ute, and Navajo nations.

For the rivers of the Southwest.
May this new world ease your flow.

CONTENTS

INTRODUCTION

Welcome from Alastair

In mid-March 2020, I was told I had COVID-19. I was probably one of the first cases in Utah. Learning I had COVID-19 was alarming for many well understood and still unknown reasons, and even to this day, new variants emerge. Thank goodness for science and vaccines—I'm vaxxed. On March 11, the World Health Organization had declared a global pandemic. America was soon in the throes of a lack of reliable tests to detect the coronavirus in humans, wishy-washy guidance from the Centers for Disease Control under the Trump administration, Utah Jazz COVID-19 diagnoses shutting down the NBA, and an earthquake on March 18 followed by hundreds of aftershocks. Then, in May, the tragic death of George Floyd inspired and re-energized racial justice uprisings and the Black Lives Matter movement, a global call to action to end police brutality toward Black, Indigenous, and People of Color communities. This was 2020, and then some.

Across Indigenous communities, the impacts of climate change, along with COVID-19, have resulted in the devastation and loss of our people. I knew more losses, and it still hurts. Our cultures, languages, and Indigenous knowledge systems, which could indeed help with solutions for the world, including the ongoing climate crisis, helped inform Indigenous leaders who directed communities to scale back travel and economic activity for the safety of all. It is often understood from Indigenous thinking that the coronavirus is an obvious indicator of the need for global change, including the dismantling of white supremacy and capitalism. This was clear as day in summer 2020. I had been walking to my home in downtown Salt Lake City during a mandatory curfew, an effort by public health officials to mitigate the virus's spread. The abandoned streets were eerie, like you see in a pandemic blockbuster on the theatre screen. But, it was also a beautiful moment to see living ecosystems displaced by concrete jungles reclaim their spaces. The impacts of the virus uprooted capitalism like the snap of a finger, revealing the need for greater change. Walking downtown that day made me realize how much harm humans have caused to Nihima Nahasdzaan, Mother Earth, and Yadilhxil Shita, Father Sky, and other non-human life.

The traumatic truths of 2020 forced me to dig deep and find pathways to help our global humanity. It took a while to get here after some hard truths like heartbreak, reconciliation, love, forgiveness, therapy, ceremony, and tons of internal healing. Some of this heartbreak happened when medical personnel had no idea how to consult me at two local health-care settings for COVID-19. Cloaked in their PPE, health-care workers stood away from me in the emergency room, as if I were something uranium. I knew I was sick, but because the virus was still new to Utah and tests were scarce, I didn't know for sure I had COVID-19. In those early stages of the pandemic, I felt the structural racism of health-care access. Even so, I did my best to advocate to get tested after doctors ruled out the common cold and flu. Through that experience, I wondered how COVID-19 would become an issue for BIPOC communities among Utah's majority white population. Fourteen days later, after many attempts for a test, I learned I was positive for COVID-19. It was a lot to process, but it helped me learn to accept what I had. I took solace in Diné healing practices, and to this day I thank my family and healers and their cultural knowledge for the ongoing healing. The reciprocal love for Mother Earth showed me the power of her gifts of sage, pine, juniper, tobacco, and other medicinal herbs. Thank you, Holy People and Creator, for my breath.

Now that I feel pretty good and healed-up, I am working to address the effects of capitalism, racism, the patriarchy, and the exploitation of our lands—current and ancestral Indigenous territories.

In June 2020, I was asked by Torrey House Press to be co-editor of this anthology with Brooke Larsen. It was mostly a shock because it was an unexpected writing opportunity. I was slowly coming out of four months of isolation and quarantine and met with Brooke and Kirsten Allen, publisher for Torrey House Press, to talk about the vision and contributors that would be part of this project at Salt Lake City's Sugar House Park. I also questioned Brooke and Kirsten about the validity of the invitation to co-edit, only because of how diversity, equity, and inclusion are becoming trending words after the onset of the pandemic and uprisings. I did not want to be tokenized as a co-editor. But, it made sense to help direct *New World Coming* because it is voices like Brooke's, the contributors', and mine that do really matter. Our voices are what the world needs right now.

I'm so appreciative of this opportunity to write and elevate voices not often heard. Brooke and I tease each other that we are essentially amplifying the voices of our friends. Even if they are our friends, we know them as frontline organizers, who stand up for diversity, equity, and inclusion in all communities across the Intermountain West and the world. Our contributors are the true stars of this anthology. We're merely helping them shine their light brighter for the world to listen and issue their calls to action to end the climate crisis.

In Diné culture, we are known to be living in the Glittering World. And just from listening to the collective knowledge of my people, I often thought about whether we—as humans and non-humans—would transition into another dimension of a New World Coming. Prophecies and Indigenous narratives say so, and I think with this anthology and the truth of the world, we are transitioning into a New World, climbing the reed just like how Turkey took seeds from the previous underworlds to sustain human nature. As they say, never ever lose hope. Sihasin. Ahehee nitsaago!

Welcome from Brooke

In April 2020, I was asked by Torrey House to edit a book that looked at the coronavirus pandemic from a climate activist perspective. A lot of my adult life has been spent in the youth climate movement, that eighteen to thirty age range of which I'm quickly approaching the upper end. During the early weeks of the pandemic, articles were circulating about what the climate movement could learn from the massive transformations happening—mutual aid networks, a global consciousness, an awakening to racial and socioeconomic injustices, new public health infrastructure, science communication, a collective slow down.

However, in those early days, we didn't quite see how after initial shutdowns and shock, baselines would shift and the thousands of daily deaths would become numbers on a screen until one of those deaths touched us personally. Just like the government "leaders" have done with the climate crisis, officials put profit over health as they failed for months to pass a relief package that made staying home and safe accessible for everyone. Billionaires continued to get richer, police continued to kill Black people, ICE continued to rip apart families at the border, and corporations continued to pillage Native land. As people waited on hold for hours trying to get an unemployment check, the connections between all these intersecting crises became impossible to ignore. So people stood up. And it became clear that this book needed to be about so much more than what the climate movement can learn from pandemic.

That's not to say, though, that the climate crisis is not a central part of this book. The climate crisis we face is a direct result of the systems of destruction, extraction, and exploitation—white supremacy, capitalism, imperialism, the cisheteropatriarchy—that have been killing people and the more-than-human world for centuries. Because of that, there are no essays solely about the climate crisis. Rather, drought and wildfire, pollution and heat waves, surface alongside conversations about migrant justice or voter suppression. That's how we experienced climate change in 2020, and that's how we'll continue to experience climate change going forward.

Throughout 2020, I felt the climate crisis in the constant scratch and tightening at the back of my throat each time I stepped outside into the smog of wildfire season. I saw it in the heat waves that radiated off the pavement. I smelled it in the smoke that was never punctured by the scent of crisp rain that usually explodes from late summer skies. I felt it on my dry, overwashed hands. I tasted it in salty sweat that dampened my mask as I chanted "Black Lives Matter" and called the names of Black men and women killed by police. I heard it in sobs piercing my car speakers as NPR broadcasted stories about loved ones lost to pandemic, a public health crisis that will only occur more frequently in a warming world.

A couple years ago, I sat around a table with other young climate activists from across the Southwest. We met in Flagstaff, Arizona, to strategize the future of Uplift, one of the region's youth-led climate justice organizations. Our throats ached as a fire burned in the mountains at the edge of town. On the last day, the facilitator asked us to imagine our movements won. What if climate justice and our intersecting dreams were achieved? What does it taste like? Smell like? Sound like? What do you see? What do you feel on your skin?

Everyone talked about water. The sound of children splashing in rain puddles. The feel of a cold river running through hot sandstone. The smell of well-watered soil. The taste of crisp, clean water and fresh veggies from abundant gardens. The sight of springs that haven't gone dry.

What does hope mean, when the most basic element of life feels in question?

In this book, we're interested in the active kind of hope that requires courage and choice, the hope that lives in the waves of resonance between my rage and your sorrow, between my joy and your celebration. Hope birthed anew as snow fell in Salt Lake City on the same day news outlets declared Trump lost the election, the same weekend Las Vegas received its first drop of rain in over 200 days. Hope rises today with our collective realization that humans imagined these systems that are killing people and the planet, and that we can imagine our way out of this mess.

Welcome from Both of Us

Throughout 2020, social justice movements declared "we can't go back to normal." Sonya Renee Taylor aptly said,

> "We will not go back to normal. Normal never was. Our pre-corona existence was never normal other than we normalized greed, inequity, exhaustion, depletion, extraction, disconnection, confusion, rage, hoarding, hate and lack. We should not long to return, my friends. We are being given the opportunity to stitch a new garment. One that fits all of humanity and nature."

This book, *New World Coming*, is a response to that resounding message. Moments of crises can lead to moments of transformation, and we asked ourselves and the book's contributors, what is the new world we want to create? The voices in the following pages provide pathways for new ways of being that values people over profit, centers community care rather than individualism, respects and nourishes the earth rather than extracting and exploiting its gifts.

Our contributors also show that as we build the new, we also carry with us the teachings of our ancestors and the lessons of history. Building new worlds is a cyclical journey. We don't believe we will ever reach an end where we have "won." We will have celebration, joy, and healing. But we will also have to continue to adapt, shift, and defend what we love and believe. Just like the moon's phases, the journey towards justice and liberation will have moments of darkness and brightness, fullness and mere slivers. We are constantly learning and birthing anew. Our history and our future sit with us in the present.

That's why we structured the book based on the moon cycles. Though all pieces carry aspects of each of these phases, the first section, New Moon, focuses on themes such as history, blood memory, ancestry, and roots. The second section, Quarter Moon, zooms in to 2020 and particular events that unfolded during that tumultuous year. The third section, Full Moon, illuminates what new stories, systems, and ways of being we are birthing.

All of the contributors are from, live in, or have ancestral connections to the Southwest. This was an intentional decision for two reasons. First, our publisher, Torrey House Press, is a nonprofit based in the Intermountain West. Because of that, we wanted to give a platform for people in our region who often don't have access to major coastal publishers or a say in how their story is shared. Second, our relationships are rooted in the Southwest. Rather than going into communities where we have no ties, we compiled this book based on trusting and reciprocal relationships we have built over years. Many of the contributors are friends, colleagues, people we heard speak at events we organized, someone we stood next to at a protest, a friend of a friend. This is a book of relationships because we will need our relationships to build a new world.

The contributors work with a variety of theories of change, strategies, and tactics. Some pieces center on anarchist and abolitionist ideologies, while other contributors discuss their experiences working in electoral politics. We think it's important to learn from these varied approaches to not only highlight different entry points and pathways, but also to embrace complexity in conversations about how change happens.

Before you dive in, we want to explain some terminology and style choices up front. BIPOC stands for Black, Indigenous, and People of Color. POC stands for People of Color. We use Diné rather than Navajo, unless referring to specific entities such as the Navajo Nation or because a contributor requested we do otherwise. When our contributors talk about abolition, they specifically refer to the abolition of carceral systems such as police and prisons, but also more broadly the abolition of ideological frameworks that justify prisons such as punishment, violence, domination, and control. Similar to how abolitionists in the 1800s didn't believe slavery could be reformed, abolitionists today don't believe we can reform the prison industrial complex to be better or just.

We feel deeply honored and grateful to serve as co-editors of this anthology. The process was humbling to say the least. We texted almost every day and were on Zoom a lot, all in between our daily hustle to pay the bills. Between messaging about an upcoming interview or edits we needed to send to writers, we'd also talk about our therapy sessions, workout routines, makeup tutorials, coffee shops, and big life decisions like where to live and how to navigate the challenges of pandemic. Before this book, we were colleagues and knew each other as fellow writers, but through the process, we became dear friends. Through the ups and downs of compiling this book, it was our friendship and the moving stories from the contributors that kept us going. We hope the stories and voices in these pages also inspire and resonate with you.

PART 1 NEW MOON

The New Moon is not visible to the human eye, yet it also marks the beginning of a new cycle. Similarly, our histories, memories, and roots are not always readily in view but the wisdom of our ancestors, impacts of generational trauma, and sacred texts continue to shape us and inform our present. In this section, contributors speak to key political and cultural moments from the past that led to the intersecting crises that ruptured in 2020. They also speak to the teachings and resilience passed down through generations with concepts such as blood memory, traditional knowledge, and faith. This begins the cycle for building a new world.

CREATION

BY LINDA HOGAN

I am a warrior
wanting this world to survive
never forgotten, this earth
which gave birth to the bison, the scissortail,
the vultures of Tibet who consume the finally released
mystics like my own old ones
who taught that we are always a breath,
a breath away from bullets.

I am from a line of songs,
a piece of history told by our people.
In every gully lies the power of a forest song waiting to begin,
the first ones sang when they crossed into this existence
and down to the canyon where I live.
I dreamed they passed
the creek-bed, each canyon wall,
the stones I love, lichens growing on them,
the route I go to the river where bear also fish.

It is hard for some to know
the world is a living being.
They live with forgotten truth
replaced with belief. Perhaps that's why
the books of the Mayans were burned,
and written languages destroyed in the North.

You can weep over such things
as lost love, or the passing of loved ones,
but always remember those birds, the bison,
their grief, too, and how the land hurts

in more chambers than one small heart
may ever hold.

Linda Hogan is a Chickasaw poet, novelist, essayist, playwright, teacher, and activist. Her poetry collections include A History of Kindness, *winner of the Colorado and Oklahoma Book Awards,* Dark. Sweet., *and others. Her fiction has garnered many honors, including a Pulitzer Prize nomination for* Mean Spirit. *Recipient of a Lannan Literary Award, a National Artist Fellowship from the Native Arts and Culture Foundation, a Lifetime Achievement Award from the Native Writers Circle of the Americas, and the PEN Thoreau Prize, she lives in Colorado.*

NOW AN ANCESTOR

LEARNING FROM THOSE WE LOST

BY BROOKE LARSEN

COVID-19 touches most pages of this book—from reflections on mutual aid and profiles on health-care workers to the intersecting crises that a pandemic made impossible to ignore. For me, though, I notice COVID's presence most strongly in the missing words.

In late October 2020, Alastair Lee Bitsóí and I were scheduled to interview Margarita Satini, a Utah-based community organizer. We hoped to include her story in this book and highlight her work on climate justice, the census, and issues impacting her Pacific Islander community. But that interview never happened. The morning we were supposed to interview her, I woke up feeling sick, so I asked to reschedule. But I'll always regret that decision, because a few days later Margarita got COVID-19, and a week later she died.

Margarita, like other leaders gone too soon, their lives cut short by preventable violence or illness, became an ancestor in 2020. Many of our readers and many of the contributors to this book also lost loved ones, and I hope whole books are written with tributes to the millions of people lost during the COVID-19 pandemic. All of their stories deserve to be shared. I decided to honor Margarita here because her death is part of *New World Coming*'s story. Her words were supposed to be here.

I've watched video tributes of others who died from COVID-19—moms dancing, grandparents laughing, college students dreaming. I have been most struck by their vitality. When Margarita passed, people kept saying she was "larger than life." She was fierce and loud. People have said that the ground literally shook when she stepped up to the mic. She brought her abundant energy not only to community organizing but also to her family. She lived in an intergenerational household with her husband, children, and grandchildren—the kind of loving home that tragically became most vulnerable to the spread of COVID-19. Her liveliness made her death seem unfathomable. I imagine many others felt that way about those they lost. I wonder if we remember the moments of liveliness most strongly to make the death feel less real.

The news articles say Margarita died from COVID-19 complications. But that would be a cruelly insufficient explanation for Margarita. She, and millions more, died because elected officials around the world failed to protect our people during a pandemic. When the governor of Utah tweeted his condolences, many of us felt not only immense grief but deep rage boiling in our bellies, since he had failed to listen to the demands of Margarita and her community time and time again.

Margarita was the chair of the Utah Pacific Islander Civic Engagement Coalition and an organizer with the Sierra Club and People's Energy Movement. In between those roles, she also spent 2020 assisting in census efforts to make sure people of color were counted.

When COVID-19 started hitting her Pacific Islander community and people of color hard, she volunteered as a community health worker. She pushed the state legislature and governor to pass rent relief and freeze utility shut offs, because she knew that economic security was necessary for people to stay home and stay safe. She also knew how best to connect with people, meeting them where they are. I'll always remember when, in the summer of 2020, Margarita live streamed herself getting a nose swab COVID-19 test to encourage her community to get tested. It was funny and serious at the same time, this powerful combination that Margarita always seemed to pull off so smoothly.

Margarita and I didn't have much one-on-one time. Our interactions were almost exclusively in organizing meetings and events we planned together. But a couple of months before she died, we were put in a Zoom breakout room together. The prompt was something about our superpower or growth edge, and the challenges we faced fully expressing our power. We both talked about how patriarchal spaces often told us we were too much, I too quiet and she too loud. In typical Margarita fashion, she made a joke and said, "I bet people just talk your ear off!" She reflected on how sometimes she felt like she should be less loud and outspoken, but she had no choice but to speak up.

In June 2020, Margarita spoke at a Black Lives Matter rally in Salt Lake City, mobilizing the crowd to take action as an uprising swept the nation. Her speech, and the responsive calls of the people gathered, may bring you back to a hot mid-summer protest in your town, sweaty hands clasping a sign as you chanted Black Lives Matter behind a face mask. Maybe you'll hear the voice of one of your community leaders. Her words here honor not only her voice, but also the chants, narratives, and calls to action so many Americans were shouting during the long, hot summer of 2020.

Margarita: Good morning, Black, Indigenous, People of Color, white allies. Good morning! I just want to say Black Lives Matter!

Crowd: Black lives matter!

Margarita: If I say Black Lives, I want you to say Matter. Black Lives!

Crowd: Matter!

Margarita: Black Lives!

Crowd: Matter!

Margarita: Black Lives!

Crowd: Matter!

Margarita: Yes they do. But this country has done a piss poor job of embracing, of lifting, of protecting Black lives. So if all lives mattered, George Floyd would be alive. If all lives matter, Breonna Taylor would be alive. If all lives matter, Ahmaud Arbery would be alive. But they're not. So we have to continue the work until it is etched in the souls of every being and it's carved into the foundation of this country that Black Lives Matter.

My name is Margarita Satini, I'm a Pacific Islander, Tongan. Let me get a hoot from my community. I was going to invite my Pacific Islander community to join me up at the front stage, but my talk ain't long, and it might take a while. So I just want my Pacific Islander AAPI people to raise their hands. I want the Black community to see that my community is here. We stand in solidarity with every single one of you. We are here to fight injustice. As a woman, as a non-Black person of color, who has experienced racism, discrimination, sexism, it has not been at the level that our Black community has faced.

If we truly value Black lives, we have to continue to work after this protest. It doesn't stop here. We got to do work after the rallies, after the marches. Who here is registered to vote? Raise your hand. Wooo, Okayyy. Say it with your chest now. All of you guys who are not registered to vote, we have several tables over here to the side. We're registering people to vote, and we're also helping people fill out their census. Who here has filled out their census? Damn, we got some woke people up in the crowd today. Yes we do.

It is our turn to mobilize, to organize, to strategize, and to build the future we want. If you're here today, it tells me that you're ready to take this fight after this protest, that you're willing to stand with all of us at the halls of the Capitol building and encourage and demand our legislators to pass bills that don't hurt the Black, Indigenous, and People of Color communities. But we need your voices. All of our white allies, raise your hands. We need you. We need your voices. We need your privilege to get us in those spaces. We need you to fight for us.

If you're here today, that tells me you're ready to take this fight to the streets. Not only are we willing to stand at the Capitol, in the halls and talk to our legislators, we are willing to stand in the city halls, talk to our mayors, talk to our city council people, and let them know no more funding the police department, divest all funds, pay for social programs, programs that strengthen the community. We don't want to invest in their policies. We ain't got time for that. Right?

I want everyone to know and I especially want the Black lives to know, we're here for you. You know this. We don't play. Black Lives Matter!

Crowd: Black Lives Matter!

Margarita: Black Lives Matter!

Crowd: Black Lives Matter!

Margarita: Black Lives Matter!

Crowd: Black Lives Matter!

Margarita: I don't want to see you just here. I need to see you at the polling station. I need to see you up the hill during legislative session. I need to see you at your city council meetings. I need you to use your voice, your privilege. Talk to your leaders, run for office, support candidates who have values that align with ours. Make sure that they are running policies that help strengthen our community. Now is the time. We ain't got time to have conversations, those conversation times are over. Now is the time to stand up and fight. Now is the time that inaction is out the window. Put your boots on, and let's go.

Thank you for this opportunity. I love my Black community. I'm thankful to my Pacific Islander brothers and sisters for being here. I love you guys. Thank you. Black Lives Matter! Thank you.

Margarita was a powerful force for justice, and like all inspiring leaders, she motivated young people. Mahider Tadesse, a senior at Salt Lake City's West High School, had been volunteering at the Utah Chapter of the Sierra Club, learning the basics of organizing from Margarita when Margarita passed. For Mahider, Margarita embodied an inspiring energy of care. "Margarita floated like a butterfly and stung like a bee," Mahider wrote in a moving tribute. "She was delicate in her care for her community, yet fierce in her advocacy as she buzzed around Utah looking for the next issue to tackle. Ultimately and untimely, she died trying to protect her community from the devastating effects of COVID, like a bee does after its last sting, its last fight of defense for her and her community."

For organizers and volunteers with the People's Energy Movement, Margarita was an extraordinary teacher who showed how care can and must turn into action. As her colleagues processed her loss, they started compiling a list, "Margarita Taught Me…" with lessons like:

Margarita taught me that love is a verb.

Margarita taught me to be brave and dream big.

Margarita taught me it's never too late to reach out and invite more people to the table.

Margarita taught me righteous rage is important fuel for movement building.

Margarita taught me there is always time to talk and share stories—that's the real work.

Margarita taught me that racial justice is climate justice.

Margarita taught me that civic engagement goes beyond voting.

Margarita taught me to show up and take charge!

Margarita taught me that speaking truth builds character.

I grieve that the new world we are building—one where everyone has quality health care and restorative rest and protection during a pandemic—wasn't here yet for Margarita and so many others whose deaths were preventable.

Margarita has lessons for us as we carry on the work she started. On February 29, 2020—a few days before COVID-19 would dramatically change day to day life in the United States—Margarita spoke at an event organized by the Utah Coalition to Keep Families Together. Community members gathered on the west side of Salt Lake City in an elementary school gymnasium to talk about stopping a proposed immigration prison in nearby Evanston, Wyoming, and to highlight the intersecting aspects of migrant justice. On the old brick walls of the school, organizers hung colorful hand-painted banners, one with hands reaching towards one another and the words "Keep Families Together." Another read "Protege Nuestra Comunidad" with pink and blue flowers circling the words. The event was grassroots and authentic, like Margarita.

That day, Margarita shared her personal story of growing up in the Fair Park neighborhood of Salt Lake City with hardworking immigrant parents who made four to seven dollars an hour at their jobs. Her family's house had no heat during cold Utah winters. One year, the Mormon Church sent a man over to install a heater. "He made us feel like dirt. He made us feel less than. He demonized us for being poor," Margarita recalled. And afterward, whenever the heater was on, the family became ill, suffering from massive headaches, nausea, and vomiting. "We were literally being poisoned by a heater installed by a man who didn't like us because we were poor, and worse—we were people of color that were poor." Power companies would regularly give her family twenty-four to forty-eight hours to pay their bills before shutting off heat and electricity. Margarita went on to join the People's Energy Movement. "I knew after I visited coal country here in Utah, listened to some of the stories shared by the coal miners, heard the stories from people who are descendants of generations upon generations of coal miners, and shared the pain that our poor communities experienced from not being able to provide for their families, that it was time for the people to democratize resources that are powered by and paid for by us."

At the end of the speech, Margarita gave a call to action: "Here we are. Let us commit ourselves to taking back our power to write rules and create the world we want to live in. Here we are."

Biography on page 263.

DZIL (STRENGTH)

BY SUNNY DOOLEY

Early morning moon-rinsed light
Gleaned starlit dawn

Here among the scented juniper and sage
I find my faint voice whispering
With quiet morning breezes
Those fine words that caused into being
My ancestors' voices

Let me trace their memory with
Coarse white cornmeal
Affirming their strength and wisdom

Let me utter faith and hope upon the
Paths of their whistled stars and I will
Turn
Knowing where I come from

Kindred and with a voice reaching
To a clarity not confined to
Here.

Sunny Dooley is Nihókáá Diiyiin Diné/Earth Surface Divine Person, living in Dédeez'á'Bighåå Ní'díshchíí' biłyiłdiz/Highrange Pinetree Glen. Her matrilineal ancestry has called this place home for the past two-hundred-plus years. She is a Diné Hozhojii Hane' storyteller, poet, and organizer of positive possibilities for true change to root.

WHEN YOU DISPLACE A PEOPLE FROM THEIR ROOTS

BY SUNNY DOOLEY

As told to *Scientific American*

When a family member dies, we the Diné, whom Spanish conquistadors named the Navajo, send a notice to our local radio station so that everyone in the community can know. Usually the reading of the death notices—the names of those who have passed on, their ages, where they lived, and the names of their matrilineal and patrilineal clans—takes no more than five minutes. It used to be very rare to hear about young people dying. But this past week, I listened to forty-five minutes of death notices on KGAK Radio AM 1330. The ages ranged from twenty-six to eighty-nine, with most of the dead having been in their thirties, forties, or fifties.

I am in shock. The virus entered our community in March, through a Nazarene Christian revival in Arizona. They brought in vanloads and busloads of people from across the Navajo Nation for the gathering; then all those vans and buses returned them to their respective communities, along with the virus. There were immediate deaths because the medical facilities were not ready for it.

I am a Diné storyteller and keeper of traditions. I live alone in a hogan, a traditional octagonal log house, in Chi Chil Tah, meaning "Where the Oaks Grow," after the Gambel oaks indigenous to this region. Officially known as Vanderwagen, the community lies twenty-three miles south of Gallup, New Mexico. The pandemic reached the area in late April. On May 1, 2020, the governor of New Mexico evoked the riot act to block off all exits into Gallup to stop the spread of the virus, and only residents could get in. The lockdown extended to May 11. It was not so bad the first week, but then we started to run out of food and water.

The groundwater in parts of Vanderwagen is naturally contaminated with arsenic and uranium; in any case, few of us have the money to drill a well. Normally, my brothers and my nephew haul water in 250-gallon tanks that are in the back of a pickup truck. At Gallup they have a high-powered well: you pay five dollars in coins, put the hose in your tank and fill it up. You haul that home, dump that into your cistern, and you have water in your house. Without access to Gallup, people began to run out of water—even as we were being told to wash our hands frequently.

My hogan has electricity but no running water. My brothers bring me water, and they put it in a seventy-five-gallon barrel. I drink that water, and I wash with it, but I also buy five gallons of water for five dollars, in case I need extra. I typically use a gallon of water a day, for everything—cooking, drinking and washing up. My great-grandmother used to say, "Don't get used to drinking water, because one of these days you're going to be fighting for it." I have learned to live on very little.

We have a lot of cancers in our community, perhaps because of the uranium. And we have many other health issues that I think make this virus so viable among us. We have a lot of diabetes and a lot of heart disease. We have alcoholism. We have high rates of suicide. We have every social ill you can think of, and COVID-19 has made these vulnerabilities more apparent. I look at it as a monster that is feasting on us—because we have built the perfect human for it to invade.

Days after Gallup reopened, I drove there to mail a letter. Every fast-food establishment—McDonald's, Kentucky Fried Chicken, Wendy's,

Burger King, Panda Express, Taco Bell, they're all located on one strip—had long, long lines of cars waiting at their drive-throughs. This in a community with such high rates of diabetes. Perhaps there wasn't any food available in the very small stores located in their communities, but I also think this pandemic has triggered a lot of emotional responses that are normally hidden. On the highway to Vanderwagen, there is a convenience store where they sell liquor. And the parking lot was completely full, everybody was just buying and buying liquor. There is a sense of anxiety and panic, but I also think that a lot of Navajo people don't know how to be with themselves, because there isn't a really good, rounded, spiritual practice to anchor them.

COVID-19 is revealing what happens when you displace a people from their roots. Take a Diné teenager. She can dress Navajo, but she has no language or culture or belief system that tells her what it means to be Diné. Her grandmother was taken away at the age of five to a BIA (Bureau of Indian Affairs) boarding school and kept there until she was eighteen. At school, they taught her that her culture and her spiritual practice were of the devil and that she needed to completely deny them. Her language was not valid: "You have a Navajo accent; you must speak English more perfectly." Same happened to the teen's mother. This was also when spankings and beatings entered Diné culture. Those kids endured those horrible ways of being disciplined in the BIA schools, and that became how they disciplined their own children. Our languages were lost, the culture and traditional practices were gone.

I meet kids like this all the time—who don't know who they are. For thirty-five years I have been trying to tell them, you come from a beautiful culture. You come from one of hundreds of peoples who were thriving in the Americas when Columbus arrived; we had viable political and economic systems that were based on spiritual practices tied to the land. Some five hundred years ago, Spanish conquistadors came up the Rio Grande into North America in search of gold. They were armed with the Doctrine of Discovery, a fearful legal argument issued by the Pope that sanctioned the colonization of non-Christian territories. Then in the mid-1800s, the pioneers came from the East Coast with their belief in Manifest Destiny, their moral right to colonize the land. As their wagons moved west, the Plains Indians were moved out and put on reservations. When your spiritual practice is based on the land you're living on, and you're being herded away from what somebody else would call her temple, or mosque, or church, or cathedral—that's the first place your spirituality is attacked.

My great-great-great-great-grandfather on my father's side was captured and taken on what we call the Long Walk to Fort Sumner. Initially about ten thousand Diné people were rounded up, and many died on that walk, which took weeks or months, depending on the route on which they were taken. They were imprisoned for four years at Fort Sumner and released in 1868. At about the same time, my great-great-great-great-grandfather on my mother's side escaped from Colonel Kit Carson at Canyon de Chelly and traveled north with his goats. He came back down to this area at just about the time my great-great-great-great-great-grandmother escaped Spanish slavery. Slavery was introduced here by the Spanish—that's never talked about. Many children born at Fort Sumner were taken into Spanish families, to be slaves.

We had the 1918–1919 influenza, one of many viruses to invade our community. Then in the 1930s there was the Great Depression. We didn't know that was happening: we did not have money, but we had wealth in the form of sheep until the government came in and killed our sheep in the Stock Reduction Program. They said the sheep were eroding the land, but I think they did it because the sheep made us self-sufficient, and they couldn't allow that. We had spiritual practices around our sheep. Every time we developed self-sufficiency and a viable spiritual practice, they destroyed it. My mother said they dug deep trenches, herded the sheep, and massacred them.

A tuberculosis epidemic in the 1940s took away my mother's parents. My great-grandmother, a healer and herbalist, had hidden my mother from the government agents who snatched Diné kids to put them into BIA boarding schools. My mother became a rancher, a prolific weaver, a beautiful woman who spoke the language. She did not speak much English. She died at ninety-six; my great-grandmother died at 104. Now, in our community in Chi Chil Tah, there are no more traditional healers; the oldest person is my great-grand-aunt, who is seventy-eight. I am the only traditional Diné storyteller.

Now that we are talking about issues of race in America, we need to also talk about the Native American peoples who were displaced. There is a reservation in upstate New York of the Iroquois people—all of twenty-one square miles. How much land were the Iroquois originally living on? Who was living in what is now Massachusetts? What about Pennsylvania? What about all the states under the umbrella of the United States? Whose land are you occupying? Abraham Lincoln ordered the massacre of thirty-eight Dakota men the day after Christmas, the same week he signed the Emancipation Proclamation, yet they call him Honest Abe. They don't talk about the dark side of things, and I think that is what COVID-19 has revealed—the dark side. We see a police officer putting his full body weight on the neck of a black man. And suddenly everybody goes, Wow! What have we evolved to?

It seems to me that COVID-19 has revealed a lot of truths, everywhere in the world. If we were ignorant of the truth, it is now revealed; if we were ignoring the truth, it is now revealed. This truth is the disparity: of health, wellbeing, and human value. And now that the truth has been revealed, what are we going to do about it?

Biography on page 21.

REMEMBERING INTO THE FUTURE

BY JADE BEGAY

In Indian Country, there is a collective experience known as blood memory. Words seem to fail explaining this phenomenon because blood memory is a feeling or a knowing, but my interpretation is that blood memory is an embodied remembrance passed down from generation to generation. Some people refer to blood memory as akin to genetic or ancestral trauma or epigenetic inheritance. More simply, we pass down in our familial lineages experiences and memories. Sometimes they are good and joyful, and sometimes they are traumatic and rooted in grief.

As the coronavirus spread in the spring of 2020, North America's Indigenous Peoples carried a unique experience of stress and fear because of this blood memory. In the eighteenth century, as European settlers sought to colonize Indigenous lands, they weaponized germs, giving blankets infected with smallpox to tribal communities to slow down Native resistance and to decimate Native populations. In addition to smallpox, measles and influenza were also brought to North America during these early centuries of colonization. It is estimated that together these diseases killed 90 percent of Native Americans.

Colonial violence has led to other public health injustices and crises within Indigenous communities. In the nineteenth century, the federal government forced Native peoples onto reservations, disenfranchising Native populations and creating, to this day, vast injustices in access to public health services. During the 1970s, the Family Planning Services and Population Research Act led to the sterilization of Native women. Between 1970 and 1977, at least 25 percent of Native American women of childbearing age were sterilized.

These historical events matter in this moment because our communities remember. What's more, our bodies and our spirits remember.

Yet despite this collective remembering, this trauma, and the anxieties that this deep-rooted grief brings up, Native Peoples organized and came together in innovative and courageous ways in 2020. Once again, we demonstrated our ability to survive and thrive in the face of uncertainty and peril.

In the Navajo Nation, one of the largest tribes in the US, the COVID-19 infection rate reached the highest per capita in the US during the summer of 2020. As the Nation responded to the exponential rates of COVID-19 cases and deaths, they had to grapple with health-care gaps that have existed for decades, if not centuries. Even before cases reached the peak, there was a lack of doctors, hospital beds, respirators, and equipment across the Navajo Nation. To be sure, this problem isn't new; the coronavirus just amplified the long and shameful history of underfunding health services in Indian Country.

While these communities struggled to respond with emergency health care, they also faced food and water shortages, highlighting yet another gap: infrastructure. In the Navajo Nation, it is estimated that one in three families haul water to their homes, and it can take multiple hours to drive to a water-filling station, that is if you have transportation. Of course, during a pandemic, access to water is essential for basic health and safety. However, strict but necessary stay-at-home orders disadvantaged Navajo families' ability to survive as hauling water became limited. This is but a microcosm of what the pandemic has brought to our attention in the last year: investment in basic infrastructure and health care has been too-long neglected, leaving us unprepared for crises such as pandemics or impacts from climate change.

Yet Indigenous communities, like Navajo, showed how community care and self-determination can provide security and solutions during times like this.

Across the Navajo Nation, mutual aid powered by community members and leaders provided Navajo and Hopi families across New Mexico, Arizona, and Utah with everything from food and water to firewood, protective personal equipment (PPE), and traditional medicine bags to support people spiritually. During the summer, I traveled from my home in Santa Fe, New Mexico, to Flagstaff, Arizona, to deliver supplies to a warehouse that served as a hub for receiving goods to be distributed to Navajo families. Inside the warehouse were all of the supplies one might need from the grocery store—boxes of baby food and diapers, women's health-care products, light bulbs and, yes, toilet paper. Goods were delivered to people in need so families could avoid travel and remain safe in their homes. On this same trip to Flagstaff, I learned that Navajo doctors were also providing mutual aid to one another and sending protective equipment to remote clinics in order to stay safe as they provided care. In 2020, so many of us saw the value and deep reciprocity that exists within mutual aid. This trip for me affirmed the power of mutual aid and community organizing and how this solution will be needed in years and generations to come.

Throughout the COVID-19 pandemic, Navajo communities used social media systems to connect community members far and wide, enabling urban family members to fill out request forms so that supplies could reach their families living in rural areas. Many Navajo people do not have access to WiFi or cellular networks, another infrastructure gap that needs to be addressed as we rebuild and recover from this pandemic.

Thanks to this community organizing, food and water was delivered safely throughout the sixteen million acres of the Navajo Nation. Connecting people with mutual aid across vast distances is no small feat and requires cultural understanding to support a community this large.

It makes sense for the help to come from within, from people who speak the traditional Diné language, know how to best reach people and collect data, and, of course, understand the environment and landscape itself.

Indigenous values are woven throughout implementation of the community care we saw during the pandemic. Elders were and are being prioritized, culture and language are being integrated and honored, and, above all, the organizers and volunteers are practicing compassion and care for the whole, instead of focusing on individualism.

Perhaps this is one of the lessons or memories that has survived throughout time to serve us as Native People again and again: caring for the whole. Recall the first weeks of the pandemic when we saw American society fall into toxic individualism as masses began to panic shop and hoard supplies, creating shortages of food and healthcare supplies across the country. But not in Indian Country. Instead, we looked around our communities and responded by identifying who in our community is most vulnerable. We thought about food security, ensuring that there was abundance for our people as we navigated shelter-in-place and lockdowns. But these questions are not new for us. Because of threats like climate change, environmental destruction, and displacement, we are often faced with questions about what will best serve our survival.

Are we overly dependent on food and materials coming from nonlocal sources? Do we have energy security in case the electrical grid is damaged by extreme weather or we cannot access fossil fuels? What are the most fundamental collective values we will draw upon in high stress moments? How do we make decisions? And how do we not turn on each other? These questions and our responses to them will continue to help us thrive.

As some nations relied on mutual aid to navigate the storm of COVID-

19, some nations relied on their sovereignty to protect their people.

The Lummi Nation, in the coastal Pacific Northwest, showed how self-determination benefits tribal communities. Over the past decade, the Lummi Nation has been developing their community health-care system in an effort to fully practice their self-determination. In 2017, the tribe adopted an Emergency Health Powers Code, which provided a framework for implementing rapid responses, and in 2018 the tribe received a grant from Indian Health Services (IHS) to support self-determination in health programs. Since 2010, the Lummi health services have raised substantial revenue by treating patients on Medicaid and Medicare as part of a third-party billing program created by President Obama. This added income has enabled more financial flexibility and health autonomy, allowing the tribe to work outside the bureaucracy of the severely underfunded Indian Health Service. Because the Lummi had the financial resources and infrastructure in place before the pandemic hit, they were able to respond quickly and effectively.

Lummi medical teams led the way in responding to COVID-19 by creating preventative measures in their community long before the federal government did. As the first US case was confirmed in Seattle, just 115 miles south of the Lummi reservation, the Lummi people quickly responded. On March 3, 2020, the tribe's leaders declared a State of Public Health Emergency, ten days before the US president declared a national state of emergency, and they turned a fitness center into a field hospital—the first in the nation—to be ready as cases emerged.

The Lummi Nation's response stands as a model for other tribal communities—all communities, in fact—for how self-determination can create meaningful infrastructure and better allocate resources.

Because the Lummi Nation is not solely reliant on federal programs for accessing emergency funds, leaders were able to act more quickly to keep their people safe. In April, many tribes worried about how they would receive funding from a stimulus bill that provided $8 billion to tribal governments. And by June 2020, as COVID-19 brutally swept through Indian Country, these stimulus dollars still hadn't been distributed to tribes and their citizens. As it historically happens, federal and state bureaucracy created barriers and slowed the distribution of "emergency" funds while tribal members faced their normal food, water, and health-care shortages. The Lummi, however, were not put into a holding pattern, nor was their response as a nation to the pandemic dependent on these funds, signaling the power in building self-governance and determination.

While these two examples illustrate the potential of nations and communities to respond to crisis, individuals have also shown what Indigenous people are capable of when we reclaim our Indigenous traditional ecological knowledge in the face of existential threats.

In northern Nevada, in Numu Territories, Autumn Harry put her passion and traditional knowledge of fishing to use during this time. "Living in a rural community, it is difficult to access healthy, nutrient-dense foods. Due to the pandemic, our nearest grocery stores are still getting ransacked and items are being hoarded, forcing our rural communities to pick from the scraps. Although I can't make monetary contributions to elders during this time, I can use my fishing skills to help put ancestral foods on the table," said Harry.

Throughout March and April, Harry fished for trout in the mornings. She would take her catch home and create sterilized and safe packages for elders, demonstrating that we as Indigenous people have knowledge useful not just in this COVID-19 crisis but for generations to come.

As 2020 progressed, in addition to evolving and expanding responses to COVID-19, Native communities also took on the responsibilities to usher in a new normal.

In May, uprisings swept through the nation in response to the murder of George Floyd by police in Minnesota. This moment was a catalyst uniting movements and communities to dismantle white supremacy, another toxic sickness that was only illuminated more by the pandemic.

By summer 2020, Native communities across the country were organizing to get out the Native vote. In the Navajo Nation, horse rides were organized to encourage those living rurally to get to the polls by any means necessary, and hotlines were created to share voting information in Navajo, Hopi, and Apache languages.

Personally, I had the distinct honor of creating the Sko Vote Den Podcast, which explored the realities of voting across Indian Country through interviews with organizers, movement leaders, journalists, and researchers. This project addressed the nuances and grey areas of the Native Vote as we dove into topics like how one can stand in their sovereignty as well as participate in the US democratic process. I learned so much about the gaps and infrastructure needs when it comes to the Native vote and data on what really matters for Native voters. Not to mention how voter suppression in Indian Country goes hand in hand with the social issues we face like poverty, lack of access to technology or energy, language barriers, and so on.

Thanks to all the efforts to get out the vote in Indian Country, we voted in record breaking numbers. In states like Arizona, Minnesota, and Wisconsin, the Native vote played a significant role in turning these once red states blue, which one could argue, made it possible for the Democrats to win the 2020 election. Let's not forget, we did this during a deadly pandemic, despite our communities being hit worst by COVID-19, and despite a long history of both voter suppression and health-care gaps. And we didn't stop there. Just a month after the elections, our people advocated on social media and organized to pressure the Biden transition team to nominate the first ever Native American to lead the Department of the Interior. Now, Rep. Deb Haaland is the first ever Native American to hold this position or hold a seat in the President's Cabinet.

My dear friend Julian Brave Nosiecat often writes about how Native peoples have already survived apocalypse: We survived germ warfare. At multiple moments in history, we've survived the US army killing our food systems: the bison, the sheep. We've survived being removed from everything we know, displacement, boarding schools, and concentration camps that were meant to exterminate us. Native peoples are more than resilient: we have overcome some of the most tremendous oppressions, many of us have healed or are healing, and in doing so, we have cultivated creativity, fearlessness, and unwavering determination. No matter what the challenge is, what the crisis is, these will be characteristics, the traits, the blood memories, we will bring into our futures to build regenerative solutions for our communities.

Jade Begay, Diné and Tesuque Pueblo, is a filmmaker, communications and narrative strategist, and Indigenous rights and climate activist. Jade has partnered with organizations like Resource Media, United Nations Universal Access Project, 350.org, Indigenous Environmental Network, Bioneers, Indigenous Climate Action, the Women's Earth and Climate Action Network, Allied Media Projects, and tribal nations from the Arctic to the Amazon to create multimedia, develop strategies, and build storytelling campaigns to mobilize and increase engagement around issues like climate change, Indigenous self-determination, environmental justice, and narrative change. Jade is the Climate Justice Campaign Director at NDN Collective.

LAND BACK, BLACK LIVES MATTER, AND POST-CAPITALISM

MOVING FORWARD TOGETHER

BY KERN COLLYMORE

2020 was a year of contradictions. Along with a raging pandemic, social justice uprisings and social distancing marked the turbulent time. Simultaneously, people from all walks of life took to the streets protesting the ills of American capitalism and white supremacy. It was the year both the far right and left voiced their disdain of the American political machine. But it wasn't just "extremists" demonstrating: soccer moms, college professors, construction workers, students, marched for justice together. While it was heartening to see so many Americans out practicing their constitutional right to protest, many people were advocating for "a return to normalcy" as they organized against lockdown restrictions, decrying their lack of access to haircuts and bars. However, in many POC communities, 2020 was the year to be heard. From the Land Back movement to criminal justice reform, everywhere one turned there was some kind of social or environmental justice protest happening. It was impossible to escape.

In the Black community, the brutal murders by police of George Floyd and Breonna Taylor added to a long list of injustices, but they also shone a light on similar inequities in other POC communities. In my small town of Gallup, New Mexico, which has a large Indigenous population, we remembered Rodney Lynch, a Native American man who died at the hands of the Gallup Police Department in 2019 after being put in a chokehold while handcuffed. As stories began to come out, we realized that many communities had their own "George Floyd" experiences.

With the dog whistles to white supremacy coming from the White House, 2020 was marked by an increase in intersectional advocacy that transcended individual minority groups. From immigration to representation, Black, brown, and Asian communities nationwide pushed back against systemic racism in their calls for justice. Erasure and division have always been some of the insidious ways that white supremacy has marginalized minority communities: "Voting rights is a Black thing," "immigration is a Hispanic thing," "pollution is an Indigenous thing." 2020 exposed this lie and showed that many of us were battling the same "things": systemic white supremacy. As an Afro-Caribbean man living in the Navajo Nation, my experience centers around a BIPOC perspective but doesn't aim to ignore similar struggles in other minority communities.

For the past ten years, I have been living and working in the Navajo Nation with my Diné partner, Janene, raising our two kids and a menagerie of animals. I started my activism work on Navajo when I learned that the state of Arizona was trying to take more water from Navajo than was allocated. After years of little to no investment in our local infrastructure, then-Senators John McCain and Jon Kyl argued that Navajos were too primitive to use all the water they had allocated so it was okay for the state to take more. This was my first experience with the water wars that have been playing out in tribal communities for hundreds of years. Since then, much of my work and lived experience has been on land, water, and food initiatives with a focus on youth and community engagement.

Most recently, I implemented a water catchment initiative, installing rainwater catchment systems and small-scale water filters for home use to ninety families across the Nation. Prior to that, I worked with the Little Colorado River Watershed Chapters Association (LCRWCA) on community-led watershed restoration initiatives that spanned thirty-five Navajo chapters and focused on grassroots engagement and community solutions. This work has shown me how important water sovereignty, food security, and land restoration are to basic human rights. Many times we had to deal with roadblocks as we tried to increase community input on the way land is managed. Often, outside companies involved in resource extraction were given a pass to use water or other resources in exchange for the tax revenue from those extraction activities in opposition to the community's wishes.

We saw something similar happen at Standing Rock when thousands of Indigenous people gathered to protest the Dakota Access Pipeline, which would transport oil through the land of the Standing Rock Reservation and across the Missouri River, an important waterway for millions of people. Standing Rock was a beacon for Indigenous people and the environmentally conscious who had been engaged in similar battles across the country, but also for all POC and those lacking access to clean drinking water. The parallel with the clean water crisis in Flint, Michigan, was obvious. Standing Rock, like the Black Lives Matter (BLM) movement, brought attention to the problems faced by Indigenous Americans and it resonated with a lot of people. Water rights has always been a pressing issue for many of the Southwest tribes, including the Navajo Nation. Before Standing Rock, there were other Indigenous-led movements against resource extraction and water pollution, from #NoKXL (another pipeline through Indigenous lands) to #Noloop202 (highway through Indigenous land). It was in supporting these struggles that many Americans began to learn about the history of these movements and people who fought extractive projects long before our current generation.

This long history of BIPOC communities fighting against white supremacy and capitalism needs to be remembered. The media narrative reporting these movements as "new" or only happening because of *(you can fill in the blank here)* works to erase history. The year 2020 saw the continuation of the fight against the enslavement of Black people and theft of Indigenous lands. From the time that Black bodies were used to build a country on land stolen from Indigenous peoples, the history of Black and Indigenous Americans have been connected. However, that history is often hidden. Only by examining and reclaiming that history, can Americans begin to see that the Black struggle for liberation and the Indigenous struggle for land reclamation is a shared fight against erasure.

Today we face an apocalyptic climate crisis, but it is no stretch to say that our BIPOC communities are still dealing with the shared trauma of experiencing several apocalypses throughout historical and ongoing colonization. People fighting against white supremacy and Eurocentric, heteronormative patriarchy is not a new phenomenon. Today, the rise of BLM and the Land Back movement carries on the long tradition of fighting against resource extraction and the commodification of land and resources. Consumerism gives power to capitalism and we have begun to think of it as an inescapable next step in human economic progress. However, when we look at the rampant killing by police of POC/non-white and non-heteronormative conforming peoples, the extreme prison sentences, the lack of safe drinking water in many poor communities, we see that these ills are not accidents, but the planned outcome in a capitalistic, white supremacist state.

We live during a time where billionaires have gotten richer during a pandemic while they employ people forced to live paycheck to paycheck, sleeping in their cars, many relying on social programs to feed themselves. 2020 was the year that growing wealth inequality was increasingly seen as an indicator of a society on the brink of collapse.

There is a theory that many civilizations collapse due to the overuse of resources, and it is this exploitation of resources that BIPOC communities are fighting against. Many Indigenous societies have practices and life ways that have allowed them to live in harmony with the lands around them. We can look at history to learn how people in the past have responded to diminishing resources. Many peoples and cultures have been mixing and trading for centuries, though it is Eurocentric heteronormative patriarchy that most of us use to look at each other today.

When we say "Black Lives Matter," it is important to recognize that Indigenous relatives are experiencing a similar impact in the fight against Eurocentric dominance. How do we move forward? Together.

Society loves to sell an image of "the collapse" as this individualistic, every man for himself, survival of the fittest narrative. However, using history as a script we see that those who come together and create community are the ones who increase their chances of survival. Everyone alive today has ancestors who have survived the collapse of a civilization. Human cooperation is the best tool for that survival. Conventional wisdom holds this image of early humans as family units that didn't care for those who weren't able to contribute. But skeletal evidence shows people being cared for by the community long after debilitating accidents. We humans have evolved to cooperate, and I have witnessed this in my own life. As the COVID-19 epidemic began to rage across the Navajo Nation and the United States at large, I saw people work to help those in need in cities and Native American communities. Groups and individuals collected and delivered handmade masks, feminine hygiene products, food, solar generators, whatever little money they had, anything and everything that could be helpful in addressing human needs resulting from this pandemic.

We have seen cooperation time and time again where BIPOC come together to strengthen their communities in the face of white oppression. Many people don't know that many early Black communities collaborated with local Native Americans. There are many stories of Black people being adopted into Indigenous tribes, stories of Indigenous communities taking in Black people, and of Black communities providing aid to relocated tribal communities. These stories predate the creation of the American empire. Many of the Pueblo fighters of the Pueblo Revolt of 1680 were mixed race. Under the leadership of Popay, the Pueblos were able to plan their revolt for eight years with his trusted generals, one of whom was a man named Dimino Narvano, a half Black, half Pueblo leader.

From Tulsa to Rosewood, many early "Black Towns" were Black and Indigenous spaces. While many of us are familiar with the Underground Railroad to Canada, most of us were not taught that the Underground Railroad also went south to places like Florida and Mexico, and out West. In Florida, the US government was busy fighting the "Seminole Wars" over tribal policies accepting Black people into Seminole society, while the Wild West is filled with stories of colorful BIPOC characters. BIPOC have been bucking against Eurocentric systems of white supremacy that allow imprisoning others or stealing land based on race since first contact.

Our histories are so fascinating, complex, and long, predating white supremacy and Eurocentric patriarchy. Before Europeans came to the Americas and Africa looking for land and labor, BIPOC have been aware of each other. While excavating shell mounds in what we call California, anthropologists have unearthed shells only found on Caribbean Islands.

What our ancestors understood, and what we need to understand, is that we can't overcome white supremacy by enacting its policies, enforcing its stereotypes, or upholding its prejudices. It is time to imagine a world where we don't see people and land as exploitable tools, where we understand that social justice and environmental justice are the same struggle, and that Black issues are Native issues, that immigration, land degradation, LGBTQ rights, and many other issues are connected in their relationship to capitalism and white supremacy. That is why it is important that BIPOC and other minority groups be the leaders of solutions that imagine the possibilities of a post-racial society.

Kern Collymore is Afro-Caribbean from Trinidad and Tobago. After graduating from Columbia University, he cofounded Sixth World Solutions and has worked on land restoration, water security, and food sovereignty for over a decade. As an activist, Kern has been on the frontlines advocating for the rights of POC, helping to facilitate teach-ins, engaging in policy advocacy, assisting with mutual aid, and creating networks of support among communities of color.

BUILDING BRIDGES THROUGH ART AND MEDICINE

BY CHIP THOMAS

As told to Alastair Lee Bitsóí and Brooke Larsen
October 2020

I am a product of the '60s. I was born in 1957, and, like now, it was a fascinating time to be alive. I'm from North Carolina originally, and I grew up during the end of formal segregation. I recall as a kid seeing an occasional sign for colored and white. I lived in an all-Black neighborhood in North Carolina and my primary school was maybe two hundred yards from my house, so I could walk to school every day. All the teachers there were African American. 1968 was an interesting year for me as a sixth grader because that was the year that the public school system in Raleigh was desegregated. Kids from Black neighborhoods were being bussed across town to white neighborhoods, and vice versa. Consequently, there was a lot of violence in the schools.

My mom was a schoolteacher. My father was a physician. He was a doctor in internal medicine, so he was an influence on my ultimate career choice. After desegregation, they were concerned about the amount of violence that was happening in the school system in Raleigh and were planning to send me to a military institute in North Carolina. Many places on the eastern seaboard like North Carolina, Virginia, and South Carolina, have a lot of military bases and a lot of military institutes that feed those military bases. It's ironic to me that my parents were concerned about violence in school but didn't have an issue with me taking part in the military industrial complex. But, as fate would have it, in the summer of '69, just before I was going to start junior high school, I attended a camp up in the mountains of North Carolina in a Quaker community. There I met some really good folks, many of whom were going to this alternative boarding school that was Quaker and it was a junior high school. Amazingly, I got to go there.

At this boarding school, I was introduced to organic gardening. I was introduced to beekeeping. I was introduced to repelling, horseback riding. I was already into camping. But this experience also emphasized building community and experiential education. It was a fascinating time to be there because I was ages twelve through fifteen, when our bodies are going through physiological changes, and I wasn't as focused on the lessons I was being exposed to in that environment at that time. It wasn't until I left that I really started absorbing some of those lessons. That was one of the influences that sent me into medicine and wanting to lead a life of service.

A few years earlier, in 1964, President Johnson launched an initiative called the Great Society, building on some policies first proposed by President Kennedy. One of the programs that came out of the Great Society was Volunteers in Service to America. Then in 1972, the National Health Service Corps was established, giving health-care professionals either loan repayment or scholarship for their medical education in return for providing health-care services in underserved communities. The government hoped that health-care professionals going into these remote areas or inner-city communities would choose to stay beyond their obligated time.

When I went to medical school, I attended a predominantly African American medical school founded by a philanthropic white family in the late 1800s for freed, enslaved people. It's called Meharry. At the time, 50 percent of Black physicians practicing in the US were graduating from Meharry. But the school also has a history of community service and physicians serving small communities that have a difficult time attracting health care. That's how I came to the Navajo Nation.

Well, there's one other part to that. There was a dear, dear friend in medical school who got to know me really well. She and her husband were also National Health Service Corps scholarship recipients and arrived on the Navajo Nation in 1986. I needed to start my obligated pay-back in 1987, so they invited me to come out to interview. This woman told her husband that she felt that this place would resonate with me, that there would be many things about it that I would appreciate between working with Native cultures and it being such a physically beautiful area. I finished my obligation in 1991, and I'm still here.

I don't have any formal art training. I think my parents, when I finished college, wanted me to get a job with some security. I had thought I would take some time off and work with the Peace Corps and try to figure out what I wanted to do with my life. At that time, I had been studying drugs and enjoyed playing music. But I didn't necessarily identify as a musician or an artist. When I was a kid during the 1960s, every week my parents would get *Life*, *Look*, and *Ebony*. *Ebony* was a big glossy coffee table magazine of Black culture, similar to *Life*. But the photo essays in those magazines, I just loved looking at the pictures and the stories that they told of people in other places that I knew nothing about. I'm an only child. I have a stammer. I'm introverted, insecure. I say all that to say that it's easy for me to kind of be alone, but through music, through drumming, I was able to find some form of expression. My mom encouraged me to do visual art as a kid. She was a public school teacher and she would make posters for her classroom using an opaque overhead projector. I would help her copy and fill in the poster. She encouraged me to make collages by cutting and pasting. That's really the only art training I've had, and I gravitated to wanting to be a visual storyteller, especially as I began understanding different kinds of activism.

On my first day at the Quaker boarding school in 1969, there was a march against the Vietnam War in Washington, DC. It was perhaps the biggest demonstration that happened against the war. This is the one where veterans who received Purple Hearts and Gold Stars or whatever took them and threw them on the Capitol steps in protest of the war. I think it was like a three-day demonstration with people camping out on the National Mall. Quakers are political in the sense that they are very aware of what's happening around the world in terms of politics, generally identify as pacifist, and are all about walking the talk, and among the group of people camping out were students and some staff from this Quaker school that I attended. My parents wouldn't let me go. But that sense of activism and questioning authority was one of the lessons that I learned there, which influences some of the art that I've been driven to do here on the Navajo Nation.

When I first came to the reservation, I built a darkroom and taught myself the process of developing black-and-white film. Everything I know in photography, I learned here. I knew that the stories I wanted to tell visually were like the stories I'd seen in *Life*, *Look*, *Ebony*. As a person of color, as a Black man in the US, the negative narrative is long-standing. I didn't just want to take pictures that perpetuated the status quo, but instead to challenge the narrative of people of color. In my goings about here on the rez and spending time with people, I was thankful for the opportunity to be in those spaces, to hear those stories, to some extent capture images that told that story, to be trusted to present that material with integrity. As I learned more about the conditions of people's lives here on the rez, I've attempted to use my art to bring attention to some of those issues, like the health and economic problems of former uranium miners and the oil workers.

Every five to six years since I've been here, I take time off to travel to a different part of the world, to get out of my routine, to see the world differently I hope, and take a break in order to think outside the box. I have to get out of the box. In early February, I was in Havana, Cuba, for two weeks, and then I went from there straight to Brazil for a month. Brazil is just amazing, well, both of those places are amazing.

Back in 2009, when I came back from a sabbatical in Brazil, I started doing the art that I'm doing now, the big-scale photos. When I came back from Cuba and Brazil this year, I returned March twenty-first, so I went straight into quarantine for two weeks. I had a plethora of creative energy, having been in such dynamic places, and I wanted to do something but didn't know what and was reluctant to try to make a statement about anything regarding the mitigation of the virus.

But a friend, Nani Chacon, who is a muralist based in Albuquerque, reached out to me and said, "Hey, now would be a good time for you to create something that speaks to the mitigation of this virus." She gave me the permission to start working on something. About the same time, I was contacted by *Art Journal Open,* an art journal based in New York City, and the editor asked if I would be interested in making a poster that speaks to this moment. This was in early April. She mentioned that at 7:00 p.m. daily, citizens were coming out and giving love to the first responders, and she wanted me to make something that people could download, maybe put in their windows, that was in solidarity with what was happening in the city. So that, too, gave me the motivation to start making posters that speak to the mitigation of the virus.

I quickly identified some of the mutual aid groups in my immediate area who were doing work and got in touch with them and said, "Hey, I'm doing a series of COVID PSA posters and wonder if I can reference your site so people will know where to get money or aid if they're interested." For a long time, I didn't get any work up outside because I was really, I mean I'm still concerned about the possibility of the transmission of the virus. It took me a while to start going out and getting work put up.

I haven't talked much about the medical element of what I'm doing and, in truth, I'm working at an outpatient facility. We don't have hospital beds and it's a small clinic so when I returned to work in early

April, most of what I was doing was telehealth, telemedicine. We have an urgent care department where suspected COVID patients are seen, but we weren't getting the numbers of patients that some of the other bigger clinics and hospitals around the nation were. We're slowly opening up. I think we're seeing 25 percent of our usual volume at this point. But it continues to be that the majority of the interactions are telehealth visits. I have to say, this is just me speaking as someone who has been in the community for thirty years, I'm devastated by all the loss and the number of friends and coworkers I've lost. I can only imagine what the families of these people are going through. It's been a challenging time.

As a Black doctor serving Indigenous communities, it's hard to see the inequities in health care, and it was really hard being here during the resurgence of the Black Lives Matter movement in the spring, going into the summer. As much as I love being here, there are still people who refer to me as colored. There is a guy I take care of in the clinic. I see both him and his wife, and one of my colleagues told me that she had a long conversation with him where he was saying he just doesn't get it, he doesn't understand the Black Lives Matter movement. He is a Trump supporter, former veteran. So as long as I've been here, I still don't speak the language. And even if I did, I feel that the sad truth is I will never be a part of the community. For as many friends as I have, I get very few invitations to go to people's houses during the holidays and people know I'm here alone.

But when I got this invitation from the editor of *Art Journal Open* to create a poster, I told her I wanted to expand this conversation about what it means to be in this moment and invite some of my favorite poets and visual artists to Zoom to meet every couple of weeks and collaborate on producing a more extensive response. So we've been working on creating both a physical zine and an online zine. Out of the eight people involved, there is one poet who is Diné, there is one poet who identifies as queer and she's Chinese, there's a Latinx man, but everyone else is African American. We became our own support group during the Black Lives Matter uprisings…I get choked up talking about this. We were able to support one another during that time.

I grew up in the '60s, during the fight for Civil Rights and against segregation. But this is the first time in my lifetime that we've had anyone in the highest office in the land literally advocate for a race war. I think it was Mos Def who made a comment recently, or it may have been Will Smith, it's not that there are more cases of the police abusing Black people, it's just there are more cameras to record the police abusing Black people. Certainly, from the time of slavery forward, the police have had a role to keep Black people in place. When people ask if it's worse now than then, I just think it's more apparent. It's not just that our capitalist economic system fosters racism, but that our economic model doesn't support people coming together and cooperating towards problem solving. That's why, during the recent uprisings, it gives me a warm feeling to see people coming together intersectionally, acknowledging problems like systemic racism in various institutions and advocating for change. Sadly, the struggle is going to continue, especially under this economic system, but I am encouraged that people are thinking in this way, appreciating how they're being affected by issues and how other groups of people are being affected, too. For me, it comes back to understanding that we have more in common than those in control would like us to see. There's lots of opportunities for intersectionality, for building across communities, across cultures. My artwork is about attempting to build bridges as opposed to building walls, about understanding environmental and racial injustices.

A population of patients I see here are men who worked in the uranium mines and mills. Hearing their stories and seeing the health consequences of their work sensitized me to this issue, to learning that here on the Navajo Nation there are over five hundred abandoned uranium sites, not just mines, but sites where there may be several mines that haven't been closed. The radioactive material leaches into the land and into the water table, into animals and ultimately into humans. In truth, it wasn't until I was introduced to this issue that I became aware of the Navajo Birth Cohort Study. It looks for impacts of uranium mining in newborns, following them for five to ten years to see how this exposure to heavy metals affects their growth and development and cognition.

For years, I lived between the Peabody Coal Mine and the Navajo Generating Station, an area rich in natural resources. The way the Nation has been exploited, it's really a case study in neocolonialism. Twenty-five, thirty percent of my patients don't have running water or electricity despite there being coal, oil, gas, natural gas, uranium, and aquifers. This lack of infrastructure in some communities on the reservation really affects them when an epidemic or a pandemic like COVID comes along. In addition to that, the exploitative extractive energy practices here on the reservation have compromised the health status of people on the reservation, which results in comorbidities that increase risk of severe disease with the coronavirus. Plus, there are a lot of multigenerational households where it's difficult to socially isolate. But Navajo Nation President Jonathan Nez said a really beautiful thing. He pointed out that this pandemic is also exploiting one of the strengths of the Navajo Nation, and that's the emphasis on family.

How do we go forward from here? I think a good starting place for everything, whether it's equitable health care, housing, education, food sovereignty, climate justice, environmental justice, racial justice, is people coming together, appreciating what we have in common in this moment and building on that. It's a real beautiful thing to see the mutual aid organizations where communities are supporting one another at a grassroots level. In that there can be a sharing and maybe some education and people building coalitions. That was the example I got from the Quaker school, where every Monday everybody in the community, maybe thirty of us, would sit in a big circle and discuss whatever. Any decision that was achieved had to be arrived at by consensus where the emphasis was on the process and building community along the way. I come back to that when I think about building stronger, more involved, sustainable communities.

Chip Thomas is a doctor of medicine at a small clinic in Dinetah (Navajo Nation). He also has a street art name, which is jetsonorama. He is a member of the Justseeds Artists Cooperative, pursuing peace through paste since 2009. Chip founded the Painted Desert Project, which connects public artists with communities through mural opportunities on the Navajo Nation.

MUSCLE MEMORY

BY MARIELLA MENDOZA

Is it muscle memory,
The way ink, when pressed against cloth at
EXACTLY the right moment, in exactly the right pressure point,
turns
into an image? The way the lines groove against surface,
The way stroke meets rhythm meets hand meets discipline?

Is it not also muscle memory,
The way my heart is used to this anguish,
The soft beats giving way to the harder ones?
When did printmaking become a cure for depression?
When did it become my coping mechanism?

You see,
For me
It's a lot like magic
Watching the images appear, with the push of my hands, my muscles
Remembering the ways to press ink against wood against cloth,
And for me
This magic is a little soothing.
Wouldn't it be nice if I could make everything appear and disappear
With a printing press?
One push
And entire worlds spill out onto canvas,
Another push
And borders disappear, cages fly open,
Comrades are freed,
And this tired muscle,
So used to caring too much, hurting too much,
Can finally stop running

-from its problems
-for its life
-from itself,
And remember what it's like,
To be still.

Mariella Mendoza is a multidisciplinary artist, writer, and media strategist with roots in the Andes and the Amazonian rainforest. Mariella's work explores their personal experiences of queerness, migration, and displacement. Mariella is currently the co-director of Uplift and a graphic design student at Salt Lake Community College.

LIVING AUTHENTICALLY
LESSONS FROM UTE CULTURE

BY BRAIDAN WEEKS

As told to Alastair Lee Bitsóí and Brooke Larsen
October 2020

I'm White River Band Ute. My family is part of the Ute Indian Tribe of the Uintah and Ouray Reservation. There are three Ute Tribes, so I do want to acknowledge the Ute Mountain Ute Tribe, which is the Weeminuche band, then the Mouache and the Capote bands of the Southern Ute Tribe as well. Collectively we make up, depending on how you're counting, six or seven bands of Ute.

I was born in the Salt Lake Valley. I went through Head Start on the Uintah and Ouray Reservation. After that, I was taken by my mom away from the rez and didn't really have contact with the Native side of my family. As I became an adult, I had the choice to return, and I reached out and started getting reconnected with family, which was a difficult situation. Some people knew me as a kid and they got excited because they knew who I was. But also, there was this hesitance of, you're not this child any longer. It's been a really great experience to work with the community and get involved with them through volunteering at first, just showing up and saying, "Hello, how can I help?"

At the time I was first reconnecting with my Ute community, I worked at a hospital pharmacy and did that for nearly ten years. The Indigenous movement to protect Bears Ears became a big deal. I realized that I'm more fulfilled in my volunteer work, so when I got offered a part-time position with Utah Diné Bikeyah I decided to test it out.

After a couple of months of encouragement to join Utah Diné Bikeyah full time I took the leap and started doing that.

It's been kind of a roller coaster of highs and lows. I don't know if this is unique to minority communities or minority activists, but at one moment you could be holding a high-high where the work is just super fulfilling to you. But also there's this low that's crushing you inside. Then you begin to wonder, is that really balance? Is that worth it? And who's going to do this work if I'm not doing it?

After I left UDB [Utah Diné Bikeyah], I went to work with an organization called Ute PAC, the political action committee for the Ute Indian Tribe, and I worked there for a year while I also worked part-time for the Ute Land Trust. Now I work full-time at the Ute Land Trust with some great people. I get to work with my tribe and my community, and the larger community that's not Native or not Ute has been receptive to doing activism or community work in a responsible and respectful way.

Throughout the hum of concern and dread about the pandemic, about COVID-19, it was hard to hear people minimizing it. The pandemic compounded everything and it manifests its way into my life. I can't sleep and there's no reason I can't sleep. I keep thinking, what projects can I do to help? How can I be successful so that my community gets the best of me or it gets the best from what we're doing together? I wake up with this anxiety that I can't explain, so I immediately just start working, and it's four o'clock in the morning, and people are wondering why they're getting emails from me. I think the dread at its core is this fear that we're going to lose people.

The majority of people who are enrolled in my tribe, the Ute Indian Tribe, are elders. If COVID-19 hit our tribe in a big way, the loss would be hard to reconcile. We talk about the loss of elders in this stratospheric way, this way that's just kind of outer space. It's very high level. For example, Mary Jane Yazzie, a Ute Mountain Ute, who's passed on now, was not just somebody I worked with, she wasn't just a board member for me, she wasn't just an elder that held space. I won't speak for all Natives, but just what I've noticed in Ute culture. These people are my friends. You start building this relationship and there are not these taboos like, oh, you can't joke around with them. They expect you to joke around with them. They expect you to feel frustrated with them or them to feel frustrated with you, to reconcile that and have a real relationship. They hold all of this knowledge and all of this culture and when they pass on, the things that they've carried on through their life go with them. If they haven't had the chance to give that to somebody, then that's a part of who we are as a people that's gone.

One of the things that I look forward to every day is our language class. We have a community language class online and many elders join. In one of our first classes, we finished our lesson and everyone's questions twenty minutes early. We have this fluent speaker who's like, "Nope, nope, nope, we got twenty minutes."

I go, okay, so we talk for like five, ten minutes more, and then she says, "Nope, nope. We still have ten minutes." She's sitting there kind of policing our time, which I kind of thought was weird because I'm thinking, well, things are done when they're done.

But at the end of the class she said, "This is my social time. I'm stuck at home. This is the time I get to speak my language. This is a time I get to share my language with you and the people get to share parts of the language I've forgotten or that I never knew about."

When I say we lose pieces of our culture, out of all these elders that join us, each one of them has language and ways to describe things that the others have forgotten or never knew. If that person were to get sick, our community as a whole would lose so much.

In my work, I get to talk with a lot of different tribal members and tribal leaders throughout the state. Across the board in Utah, from the Navajo Nation who got hit really hard to the Ute community that was able to limit exposure early on in the pandemic, everyone is asking, how do the actions that I take as a tribe or as a tribal leader or as a community leader affect not only my community and my people, but the people that are our neighbors? That kind of thinking helped a lot of the tribes. Southern Ute and Ute Mountain Ute didn't have a single case for a long time. Ute Mountain Ute ensured that the people who work at their casino, even if they're not tribal members, were paid when the casino was shut down. The Paiute Indian Tribe runs clinics in the largest geographic area in the state. They did that for Native and non-Native alike and were able to keep cases out for a long time.

Our tribe got our first case in July, and that was devastating. For the different Ute communities in the Northern Ute rez, this was serious. They were thinking about other people through this, and so was our tribal leadership. The Ute Tribe in all of its facets, the leadership, the different departments, have adapted so well to make sure that first and foremost, our community is allowed to stay healthy and happy. These cases just pop up randomly so we're not having large community contact within the rez, largely due to our tribal leadership and the work our tribal departments are doing to keep people safe. Even the Environmental Health Department, their thought is how do we take care of the environment and take care of our seniors? Even though care of seniors is not directly involved in their department, they're thinking holistically. I think that's a big deal. The tribes in Utah, not just the Ute Indian Tribe, should be commended on that holistic approach because it helped protect so many of our people and continues to do so. Tribal leadership in all of its forms have really protected our people.

That's not without us having a sense of disconnectedness from our community. The reason that some of our elders who participate in the language class are so eager to be there is that they're isolated and they're lonely, because, unfortunately, them staying at home is a necessity. Us staying at home is a necessity. It's difficult for our community, which is so interconnected. We're used to seeing each other, being able to stop at each other's house, and say, "Hey, what's up?" Or just have that doorway conversation where you come by because you told somebody you'd drop off some food, and then you're walking out the door and say "See you later" for three hours. We don't have that anymore. That's really hard. Overall, even people who kind of bicker with each other, their thoughts, first and foremost, are of us as a Ute people, as a whole. If someone is suffering in silence, I've noticed a lot of our community has taken that as, well, I need to reach out to this person. Even if they tell me they're okay, I need to push a little bit, even though as a younger person I shouldn't be pushing on an elder this hard, but I need to do that because they're trying to think of me as well.

Normally you're able to walk right into the tribal offices and see the business committee, the tribe's leaders, or see a department head or a department worker; the people who are leading programs in our community are very accessible. Now the doors are locked. You have to make appointments. All of this interaction is done at arm's length, which is a different way to think. The tribe has had to rethink the way to run our charter school, the jailhouse, the stores, and the enterprises in the Nation. How do we succeed in taking care of our community and our way of life while also succeeding in a capitalistic environment?

The Ute Land Trust is always thinking about that question. We're a nonprofit that is very directly, by design, responsive to our community, and tribal leadership acts as our board of directors. These programs that we run, like the language program, started because we had a community member reach out and say, "Hey, I want to have conversations with people around Ute language because I feel disconnected and I don't understand or speak Ute fluently, but I understand a little and I can share what I have."

The response to that, of course, is, "Well, how do we support you? Because this is your idea." They often don't know exactly what kind of support they need so we go to our volunteers, to our community, and our elders and ask their advice.

Bringing all those ideas together makes it so that we're able to connect people during COVID in a way that they feel is fulfilling and not in a way that's lip service to funders or that type of thing. The genuine concern about how to hold on to our language made all these elders and these fluent speakers say, "We'll help you. We don't want to be the face. We don't want to be recorded, but we'll help you. Tell us what you're trying to learn and we'll teach it to you." That's been great to see.

The way that we understand and perceive the world is entirely embedded in our living language. If you hear these elders talk about our language dying, they are talking about something dying that is tangible, that is living, that they're intimate with. They mourn that just as they would mourn a person. If we're going to do this work in an authentic way, we have to hold on to our language, we have to understand our language, because how are we going to service Ute culture and the Ute community in preserving our traditional knowledge or our traditional spaces if we don't understand how to do that in the correct way? Through this language class, we've learned what specific land areas meant to us, the word behind that land, the word that we have a relationship with for a specific space.

The Ute Land Trust has a few irons in the fire with easements, whether that's here in Utah, Colorado, southern Wyoming, or northern New Mexico—the places that we're from originally. We'll have people who reach out from a nonprofit in Colorado and say, "Hey, we know this is Ute land," but their concept of a Ute person is basically zero or a misconception.

During the last few years, the Ute Tribe has been involved with some land transfer projects, and recently three acres were transferred back to the Ute Tribe. On social media, Land Back is a hashtag and there are national campaigns, but, for us, there's some caution when it comes to anything in media or social media. If you're doing something for a genuine reason and not for recognition, things like boasting or talking about yourself are very contrary to people's sense of morality. There's a fine line to walk, where you're going to need funding from grantors or foundations or individual donors, but also need to work in a way that doesn't exploit yourself or your community's stories or beliefs or heritage.

The Land Back movement may not speak to every space that we're trying to work, because as I talk to different leaders or different elders, there's an understanding that you have to deal with the world how it is today. White Rivers and Uncompahgre Bands are in Uintah Ancestral Territory right now, or Uintah traditional space. But many people from those bands, especially my age, don't have a physical relationship with our ancestral land in Colorado. So what does that mean? When you say Land Back, does it mean, oh, all you white people need to return everything that's there and give it back to us and that's it? Or is that concept evolving? Sure, if you have the resources or you have the ability to return land to a tribe, more power to you. But if you're an average person trying to make ends meet, and you kind of have that guilt hanging over you when you hear Land Back, I don't want you to get defensive or shut down that conversation. It's about intent. It's about what you want from that. If you feel that you need to reconcile that relationship, reach out to us, reach out to the community that you're near, the tribe that you're near, because Land Back for me doesn't need to be actually giving physical land back. It could just be an acknowledgment where you change the way you think and look at the world and the way that you act with Native people. So when I walk into Meeker, where the Meeker massacre happened, they should acknowledge us in a real way. If that's giving your money, great, give us some money, help us with these programs to preserve traditional knowledge.

But to jump on to something that's hyped up is scary because you have to think, what does that concept actually look like? How does it fit with our values?

I think the reason we're received in some spaces so well is that we are authentic. We don't make excuses about what we can or cannot do. With our programs, whether that's three acres in Colorado that were transferred over to the tribe, or easements, or gaining education access, above and beyond any of the material things, most importantly we're authentic to who we are. I get this a lot from Ute people: don't focus on the material, that doesn't matter. Our hair may change, our clothes may change, but who we are as a people needs to be in every single project we do and has to be the route that we take. That is essentially the root of my work. You see that in the language program. You see that in us protecting land or protecting space or advocating for the return of traditional knowledge, like contraceptive methods or rights to even gender identity, or whether being gay is perceived as one way or the other. All of these concepts or any of these projects have to be rooted in a core Ute identity for me.

Ute people grow up with an expectation that things in life are not going to be easy all the time and life is hard and you have to meet that hardness with a little bit of hardness sometimes. Movements like Keep It In the Ground or ideas about a just transition off of oil are complicated when oil has been an important part of the Ute Tribe's economy. When I was first reconnecting with my Ute community, I didn't understand why our tribe invested so heavily in oil or why the Southern Utes did. What helped me start understanding that more was hearing an interview with Chairman Frost from the early 2000s about his view: that there is oil throughout all of our reservation, and we only drill up this portion of it. The thought is that to protect the rest, we do have to sacrifice this part, and the money that you get from oil is, unfortunately, connected to our ability to ensure that we have rights as citizens, even basic voting rights. In Utah, the year before we were allowed to vote here officially, a Supreme Court ruling said if we lived on the reservation, we were not residents of Utah and we didn't have rights. And Supreme Court cases at the federal level ruled that the Native people were not citizens of the United States. Our ability to fight those unjust laws depends on our tribe, and our tribe can only do that with money.

In the Uintah Basin, you hear about the Ute Indian Tribe, they're such strong oil producers. But if you look at corporations in the area, for every oil well we have, they have multiple. We're not just, "drill, baby, drill" as much as we possibly can. To take that right away from us because we're an easier target than big oil really isn't environmental justice. That is taking away from and subjugating another minority for the services that you enjoy, like the asphalt on roads, the gas in your car tank, the electricity that you use. For environmental justice, there's a level of sacrifice that we've already reconciled ourselves with because we have to protect ourselves in this world. There is an intent to diversify, to separate ourselves from dependence on oil money. I see the Ute Tribal Enterprise, or Ute Bison, or the gas stations, convenience stores, and restaurants as attempts to diversify our economy so that we don't have to rely so heavily on oil and gas.

I once had an elder talk to me about growing up in the Uintah Basin knowing the plants, the berries, the roots, the flowers, whatever you could utilize out there as far as food. She talks about taking a sled to the springs in the winter because she had to pull her own water since she didn't have running water. I know the perception of Northern Ute people is that we are well off and well established, but that's not how we were a generation ago, and even now, there are people on the Ute reservation who go without. Within this elder's lifetime, she went from being this knowledgeable, productive youth to essentially being a lawbreaker, because homeowners put up fences and started putting in apple orchards or whatever for their harvest, so she went from being food rich to being food poor almost overnight. She told me stories where she had to crawl under fences and steal apples that had fallen off the tree to feed herself and her siblings, when previously she and her family had plenty.

Thinking of that, somebody getting in trouble for taking apples that nobody was going to use anyway because of this push to constantly produce for a dollar, that's frustrating for me. Our Ute voices are important for understanding our stories and how Ute knowledge can affect our future. There are non-Native people who come from those generations that took so much from us, who see nothing wrong with what they did. They're trying to push their idea of a correct way of living or being that is unsustainable. We will reach a maximum capacity of being able to make things, and then what do we do? We get so concerned about the stock market if it's not growing anymore. And that's bad. What's wrong with a simple life? Even in the work I'm doing for a nonprofit trying to hold on to traditional knowledge or trying to protect spaces or land that we need access to, I'm getting told by elders, slow down, what's happening in the world right now, COVID and stuff, you need to take care of yourself. And even before that, it was slow down, you need to take care of yourself. Life is here to be; you're supposed to be happy and at peace with these things.

We were all forced with COVID to take a breath. A sense of constant production, of constantly moving forward or bettering yourself in a way that is material rather than emotional or spiritual, is so deeply rooted in American and Western culture that it hurts people to think of a life outside of that. When you tell people to stop and stay home for a while and just exist, there's this painful, painful thing that happens mentally and physically. All of that, whether it's emotion, mind, or spirit, will manifest physically for you. This pandemic was probably good for me in the long run, even though it sucks.

What I lean on is those teachings from my community. Things like, are you getting up before sunrise? You need to be doing that and giving thanks for that, and you need to be taking time every day to just exist.

Don't be watching TV. Don't be doing something else. I get in trouble with my grandma. I don't sit with her, obviously, during COVID, but when I did, I'd be weaving or beading or writing, and she would put her hand over whatever I was doing and remind me to just sit, just be. We need to find that space with ourselves. That's a lesson that a lot of people could take from COVID. How do you just sit and exist with yourself? And if you can't, what do you need to fix?

When I think about the future, there's this space that we find ourselves in as Native people, where we represent this living, breathing culture that's going to change and evolve. How do we do that in a way that's not just allowing ourselves to be colonized or doing things in a Western way and accepting all of the Western values for good or for worse, but in a way that's authentic to ourselves, that isn't leaving us stuck in stereotypes, in a cowboys-and-Indians-living-in-teepees space? I saw a photograph of this community garden from when the Uncompahgre and the White Rivers and the Uintah were first on the same reservation at the same time. People from the different bands came together, and they would give out produce to the community. I found this after we had started talking about doing community gardens in each of our areas. And I thought, gosh, we were thinking this was a new, great, wonderful idea, but the reality is, our communities had already been doing this and something stopped them from being able to do it, and we're just bringing that back. That tradition of providing for our community is something that we can hold on to and modernize. But how we do a garden is going to be different.

Creating a sustainable future means empowering communities to take care of themselves and reassess that value of always having to produce. Why do you need five hundred pounds of tomatoes? You don't. My community needs a hundred pounds of tomatoes and we do that by growing and sharing it together and that's great. That allows so much to happen, so much connection, so much preservation of our values of taking care of one another as Ute people. We go from an experience of exploitation to one of authentically living with each other. That's important, to live authentically to yourself in a way that supports your community. And by community, I also mean non-human communities: the environment and the animals and the plants and everything.

Braidan Weeks is the executive director of the Ute Land Trust, which assists in the healing of the deep wounds left by the injustice of the violent removal of the Ute Indian Tribe from ancestral lands in Utah, Colorado, New Mexico, and Arizona. The trust's work reconnects the people of these lands by engaging with other tribes and federal, state, and local governments to partner in land stewardship and traditional conservation efforts. Braidan is White River Band Ute, Nuhnahmuh.

GOING AROUND THE POTHOLES

BY LAURA TOHE

I grew up on the largest Indigenous American homeland in the USA, which we call Diné Bikéyah, also known as the Navajo Nation, that lies in parts of Arizona, New Mexico, and southern Utah. Four mountains mark each of the cardinal directions, including Dibé Nitsaa or Big Sheep Mountain, in southern Colorado. The small, isolated community of Crystal, New Mexico, where I grew up is closest to the northern mountain and had a small population of approximately three hundred people. Crystal is located at the base of the Ch'ooshgai Mountains and used to be accessible only by an unpaved road that led north and south with smaller dirt roads that branched outward from it. What it lacked from the outside world like television, newspapers, and ample telephones, was more than made up for with the lush natural beauty that surrounded Crystal. Red sandstone cliffs lined both sides of the main road, and in the summer, Rocky Mountain bee plants grew in fields that displayed their delicate lavender-colored flowers while their branches dripped with slender green pods. Yellow sunflowers opened their blossoms in the open fields where horses sometimes grazed. When it rained, an enchanting scent of moisture arose from the sandy red earth that made me want to eat it. A small stream trickled from Ch'ooshgai mountain where tiny frogs with long legs and webbed feet played. My younger brother and I took turns placing them near the edge of the foot bridge and watched them hop unwittingly into the air and drop six feet into the stream where they made a splash like little pebbles. Water was abundant with lakes only a short drive away, and aspen trees grew in the higher elevation north of Crystal. When we left home for the day or longer, we never locked the doors because we knew our possessions would be safe. Our two-bedroom duplex housed six of us, my single mother and five siblings. Next door lived my best friend, Wilma, who was Cherokee Choctaw from Oklahoma. We explored the forest and even buried a pet dog and lamb there. My brother staked a cross to mark their graves. Crystal was our world within another world that was a safe and a fun place to grow up in. I'm grateful that my mother brought us there when she found a job as the head cook at the boarding school.

The oldest building was the trading post. The white trading post men were different from the temporary outsiders. They were independent men who built one-room trading posts all over the Navajo Nation in the nineteenth and twentieth centuries and sold goods the Diné would eventually come to need and want. Among the foods the traders introduced were canned foods, sugary soda drinks, candy and processed meats, and white flour, all foods that cost more than in the towns. On my mother's paydays, we drove the hundred miles round trip to the nearest border town to buy groceries that would have to last until her next paycheck. We lived in a food desert. When our food supply ran low, my mother sometimes sneaked leftovers from the boarding school kitchen for us and our dog.

With the government boarding school came a dormitory and a community chapter house. Early on, the government schools worked with Christian groups like the Catholics who introduced a western diet in the boarding schools. Fish was served on Fridays and dairy was part of every meal, even though many Native people are lactose intolerant, including myself. These foods are foreign to our Indigenous diet of wild game and native agricultural plants. Earlier, foods such as pork, white flour, coffee, and potatoes were distributed to the Diné during their incarceration at Fort Sumner in the 1860s. The people were starving, and the women didn't know how to cook or prepare these foods. Nonetheless, we eventually came to adapt these foods after the return from Fort Sumner.

Then came the churches. The Mormon Church was run by young

men barely out of high school who stayed a year or two to fulfill their required church mission before heading back to Utah to college, work, or marriage. One group of Baptists came with a mobile home that the preacher parked across from the vacant building that became his church. The Navajo Bible Church was managed by Miss Gray a single, stern white woman who found agreement with solitary life. She left only to buy supplies and to visit her family. Past middle age with salt and pepper hair, she had an unremarkable face and was an outsider like the teachers who arrived at Crystal's doorstep. Living in Crystal proved difficult for them, and after a year or two they loaded up their U-Haul trailers and headed back to the Midwest, the South, and the places from which they came. Miss Gray was the only outsider to put her roots down in Crystal, although they didn't grow too deep. She lived in a one-room log cabin adjacent to the little church before turning it over to a Navajo preacher who led the congregation on Sundays. Everything was dark brown in her house except for the quilts and brightly colored afghans on the bed and furniture, and her house always smelled like she had baked that afternoon. My mother was gifted with hands that created all things made with yarn, thread, string, and fabric. In Miss Gray, my mother had found a kindred spirit who enjoyed sewing quilts and knitting, but we didn't attend her church. A couple of times I was sent to the log cabin alone on an errand. Though she was friendly with my mother, Miss Gray didn't like children.

On Sunday mornings the small Navajo congregation gathered for services at the Navajo Bible Church. It was the only church where the sermons and hymns were in Navajo. The preacher's voice carried throughout the little valley. Every once in a while, he punctuated his sermons with "Diyin Gaad." Then the chorus would start up with vigor as Navajo words drifted over the school grounds where their children attended and Navajo language erasure was enforced, and we were punished for speaking our mother tongue. Once scrubbed from children's mouths, the Navajo language rose again as each note reclaimed the stolen words. They were singing the language back one word at a time. It was Navajo only on Sunday mornings at Crystal and no one was in trouble for it. It seemed so natural, so ordinary and comfortable not to be caught in the liminal space between the Navajo and English languages. Diné Bizaad was again victorious as it was when the Code Talkers developed an unbreakable code in Navajo that helped save America during WWII. Whether Miss Gray intended it or not, the Navajo Bible Church created a space for our language on Sunday mornings.

During this pandemic there is even greater concern for the future of the Navajo language. UNESCO's (United Nations Educational, Scientific and Cultural Organization) website has designated the Navajo language as "vulnerable," which is defined as "most children speak the language, but it may be restricted to certain domains (e.g., home)." This might not be completely accurate because English has become the first language for many of the Navajo youth. The COVID-19 pandemic has not only affected the health of tribal nations, but it also affects the health of native languages. What does this pandemic mean for the future of the Navajo language, the Navajo people, and the future generations as the virus continues to take numbers of the older generation who still speak Navajo as their first language and practice the spiritual teachings and beliefs of Diné culture? Will there be a gap that further erodes our language and cultural knowledge base after the pandemic is under control? Without our language what repercussions does it have for our spiritual practices and connections to the earth, environment, biodiversity, and to each other? We have said that we are a nation who can adapt to change and are resilient.

When I used to visit my late brother in another small community, much like Crystal, I had to take a short drive off the paved road onto a dirt road that had developed deep grooves after many seasons of rain, snow, and wind. When this happens, people simply drive around the grooves and potholes. Eventually a new side road is made, which is what happened near the cattle guard where a pothole appeared. I didn't know how deep it was because it was full of rainwater, and when I drove into it, the pothole nearly swallowed the tire and could have damaged my car. Then I noticed the side road. People had adapted to the pothole by simply driving around it, because we know our communities don't have the infrastructure to repair or pave the roads. My late uncle who had worked as a community developer remarked that because the community had a chapter house, post office, and a large housing neighborhood, and since children rode the school bus every day, the road should be paved for safety reasons. After I broke away from my "go around the pothole" thinking and acquiescing to "that's just the way things are," I saw his point that having good, maintained roads contributes to public safety and a sense of well-being. And something should be done about it.

Can we adapt by simply "going around the potholes" created by COVID-19? We are living in a different time from my youth in Crystal when the Navajo homeland was safer, and we didn't have many of the social ills we have now. While the duplex we lived in at Crystal had all the modern conveniences, my paternal grandmother's hogan didn't. Water had to be hauled from the windmill, wood and coal had to be chopped and brought in, all from a distance. A kerosene lamp lit her hogan. We ate a lot of canned foods because she didn't have a refrigerator. All of the conditions my grandmother and I grew up in—food deserts and a diet that didn't agree with my body, life in compact living spaces, etc. are still the conditions that many Diné still live under, especially the elders. Such conditions helped spread COVID-19 quickly. While the pandemic revealed the deficiencies in our communities, it also revealed the work we must do to improve our lives.

We must start to envision how we can improve our communities and how we can rebuild and reclaim our language and our food sovereignty. Growing our own Indigenous food can help control the rates of hypertension, diabetes, obesity, heart disease, and asthma, diseases that appeared with food deserts, boarding school diets, fast food, and when we stopped planting our cornfields. There is much work to be done but we can begin to take the steps toward improving our health and health-care systems by calling on our tribal leaders, locally and nationally, to help bring about a better infrastructure. Communities need closer access to healthy plant-based foods rather than just processed foods that will continue to damage our bodies and may eventually lead to taking medications that maintain illness rather than heal. Incorporating exercise programs into our lives, such as walking and running, is part of participating in our own wellness, which is part of our cultural teachings, especially for the youth. I admire that my great-grandmother in her eighties still hitchhiked to the nearest border town. Herding sheep on a daily basis was how she and people of her generation took their exercise. We think she lived to approximately one hundred years of age; Diné people from her generation didn't keep track of their birthdates.

Living in the time of the COVID-19 pandemic has revealed the enormous deficits in our communities that we have lived with since settler colonialism. Once invisible in mainstream news reportage, the media exposed the poverty, the lack of nearby medical resources, lack of healthy water, food deserts, over-crowded living spaces, absence of reliable internet service that has taken its toll on schools, and the underlying health conditions that make the young and old vulnerable to the virus in in one of the wealthiest and most technologically advanced countries in the world. We say we are a resilient people and have adapted to many changes that came from the western world. We must find better ways to adapt rather than just going around the potholes. COVID-19 tells us to regenerate, re-envision, and reimagine ourselves. This is an opportunity for us to restart and abandon old systems and ways that did not work for us. Creating a food sovereignty movement by growing our own Indigenous foods that agree with and nourish our bodies can also help maintain the health of native languages. Embedded in our native languages we find the critical knowledge that sustains food sovereignty. I have often heard the Navajo expression that achieving a goal is up to the individual. T'áánihí 'adéét'įįgo t'ééyá means it is up to us to imagine a regenerative future with greater possibilities. The COVID-19 pandemic taught us that we can no longer afford to go around the potholes. It is up to us to create a well-maintained road that leads toward our sovereignty in how we choose to strengthen and better our communities, to improve the quality of our lives, and, just as importantly, for the benefit of the future generations.

Laura Tohe is Diné and the current Navajo Nation Poet Laureate and is Professor Emeritus with Distinction from Arizona State University. She has published five books and two librettos that world-premiered in Arizona and France. Among her awards are the 2020 Academy of American Poetry Fellowship and the 2019 American Indian Festival of Writers Award.

RECONNECTING WITH MY MUSLIM FAITH THROUGH ORGANIZING

BY MISHKA BANURI

As told to Alastair Lee Bitsóí and Brooke Larsen
November 2020

I don't really know when I started organizing or activism. I always tell people that I started in seventh grade with a small project to raise awareness about Malala Yousafzai, the Pakistani activist who was shot by the Taliban for trying to advocate for rights for women and girls to get equal education. I remember that being a very central conversation at the dinner table. Then I remember talking about it in my school, and no one knew who she was. I felt like it was important to talk about inequity in terms of education and gender. I started that kind of mini project, and kept going with different projects, and my mom was a huge supporter in that. I think she really pushed me to get more involved and to think outside the box in terms of what could be done. My mom is someone who I've learned from and who has taught me along the way.

I started climate organizing in my freshman year of high school. We, the Environmental Club, went on a camping trip to Bears Ears National Monument and made a video to send to President Obama asking him to make it into an Indigenous-led national monument.

I was always interested in doing environmental work, but when I was a sophomore in high school, I got into a car accident and was out of school for a while. I couldn't do my assignments in the way that I used to be able to do. I wasn't allowed to go on my laptop or read or stimulate my brain in the same ways. Because I had all this time, I got involved with the People's Climate March organizing in Utah. I think that became a way for me to deal with the struggles that I was having on my own, feeling very alone. Once I started organizing, I felt like I was in a community of people that understood what my abilities were, and that I was contributing to something.

At the time, I was struggling with my faith. I consumed media that taught me that Islam is anti-women, anti-queer, and I didn't want to identify with either of those. I struggled to identify as Muslim. But the more I organized around environmental issues, the more I learned just how in tune with my faith my organizing was, especially around environmental issues. Organizing was a way for me to reconnect with my faith and practice my faith in ways that I never thought I could before.

I got involved with the effort to pass the Utah Climate Resolution for a couple of years, and in that time I saw a lot of youth engagement and desire from youth to engage in the environmental movement. So I organized, with a couple of other people, the first Utah Youth Environmental Summit for young people to get trained and learn about environmental issues happening in Utah.

I was kind of thrown into the deep end of organizing, especially in Utah, and there's a lot of reasons for that. I think the Utah People's Climate March was the first big environmental event in Utah to talk about intersectionality. I was grateful to be part of that, but I also feel like some aspects of it were tokenizing. I've been thinking a lot about ways that we ask young people to be the hope or give us hope. I wonder about media outlets that look to young people for hope, what support do they give to young people or to retired people who are also doing a lot of work on climate organizing? I've heard people question what the community provides young people in terms of housing, funding, and support for other problems in their lives that aren't just environmental, which resonates with me. How are we supporting organizers in ways that allow them to show up fully? Maybe we can build capacity and remove barriers for more people to engage. As a young organizer, I felt really burned out really, really quickly. After I graduated high school, I took a gap year. I was too tired to do a lot of stuff, and I still am. I've been trying to balance organizing with my own life in a healthier way. I think that's something that not a lot of people go through at such a young age. I understand myself a little bit better than I think a lot of my peers do in some ways, and my faith is a part of that.

In 2013, I was at the Parliament of World Religions, an interfaith conference that met in Salt Lake City, and there was a plenary on climate change. The obvious connection they made was that the earth is one thing that we all have in common, and no matter what faith we are or what our cultural background is, something that we can all come together and fight for is the earth. That was the first connection I made between faith and the environment, hearing Muslim speakers talk about how everything in Islam is very in tune with nature because that's something that God created for all of us to practice with. The way that we pray is according to where the sun is. There's a lot of holidays that are based off observations of the moon. Water is really important. Something that I'm reminded of constantly by my parents is this story of how different beings are created. Angels are made of light. Jinns, which are like spirits, are made out of fire. And humans are made out of clay. We're literally made out of the earth. There is no separation between people and the earth, and if the earth isn't healthy, then we will not be healthy. We see that very clearly now in drastic climate impacts.

One of the ways that my faith strengthens me is by listening to the history of the Prophet and other historical figures fighting oppression and how integral anti-oppression is to just being Muslim. For me, a lot of climate organizing and fighting for the environment is being in tune with that origin and history of Islam in fighting injustice. I've been studying the Koran more. The number of verses that talk about protecting the earth because it's something that God gave us to protect for future generations surprised me, because that's not something that we learn in school or in Islamic School. We don't learn how important the earth is. That's not a mainstream thing that we talk about, which shocks me. It's important to learn about, and it feels reaffirming that I'm doing the right thing when I see environmental justice in my faith. The idea of protecting land is something that I always thought was important, but the fact that it is in a holy book of ours confirms that I am doing the work that I'm put on earth for. That's one of the reasons why I connected with my faith more, doing the research and seeing that the discussion about protecting the earth is also in the Koran. My idea of Islam isn't wrong, though I always thought it was because of the media that I was consuming and what other people were telling me about Islam.

When I was younger, I would tweet about Islamophobia and was shocked at the sheer amount of trolls that would come and comment the most dehumanizing things. It was hard for me to feel like I have to justify my humanity and my goodness, especially as a teen. But what really made it hard for me and my faith was liberal Islamophobia—for example, Bill Maher, the TV show host. It's his favorite topic to talk about, Islam, and how awful it is for women and how badly Muslim countries treat queer folks, to show that Islam is evil. I always grew up being feminist, so hearing these things and the applause in the audience when people would talk about Islam in certain ways, I felt like, I don't want to be antiwomen or antiqueer, that's not who I am, so at a very internal, deep level, I rejected Islam. But my family was Muslim, so I knew we weren't bad people. I would defend it outwardly, while internally, I would reject it.

Through organizing, I've been given the tools to understand just how dangerous rhetoric like Maher's is in terms of how it translates into policy and actions implemented by Western countries to justify imperialism. What a just world looks like to me is dismantling those narratives around Islam. I feel like everyone, even the Muslim community, treats Muslims in the US like a monolith, that we all believe the same things, that we all practice the same way. Over the years, I've learned that's just not true. People practice in different ways, and we have to be able to accept that. I've been more drawn to justice as a core value of Islam and reframing everything with that. A just world includes centering justice in Islam and in the way that we view Muslims globally.

I'm a first-generation Pakistani American, so I'm interested in the intersection of climate issues around the world. Going to school in Vermont is weird because I'm in this totally new community, but I feel so connected to the one in Utah. This separation seems similar to what it's like being an immigrant. There is this place that I have never been to, but that I love because it's what created my parents and shaped them and created our culture. Having this understanding that there are people you love across the world who are facing some real issues, like the impacts of climate, of colonization and injustice done by the Western world, that always gave me a global lens. At the same time, I was an organizer who focused on issues in Utah, and I worked hard to connect locally.

One way that there are intersections between Islamophobia and climate is in rhetoric about the Arab world or Muslims right now. We equate being gay friendly to modern, and then we separate the Western world as gay friendly, therefore it is modern and the Arab world or other worlds are not gay friendly, and therefore they're barbaric and they don't deserve to have their own state or we need to save the people that are there. This narrative justifies intervention and imperialism. And a lot of the times, like all the time, the intervention is not to save people. Rather, it's for oil or resources. Before the Iraq War, oil in Iraq was nationalized so the government could benefit from it. After the war, it was privatized. Specific oil interests were going after Iraq for a very specific reason, but it was under the guise of protection of American values, which are supposedly gay friendly or feminist, but we know that isn't the reality in our country. Also there's documents showing the military referring to the Middle East as Indian Territory. Obviously, the US has pillaged and extracted a lot on Indigenous lands at the expense of the people, and this equating of these two communities or places shows how the US views these others that they put into this box. That's one of the ways I've been trying to get the Muslim community to be in solidarity with other communities, because the way that the US military views us is very similar to how the US government has viewed Native Americans. The industrial or imperialist goal whether in the Middle East or in America is really the same: money and capital.

My organizing has helped me understand a lot about myself and how to make a difference, and my family has, too. My mom always told me how engaged she was in her community when she was younger, and I saw that more when I was growing up. She was always volunteering and providing support for people. I was the only daughter—I have two brothers—so I think because of that, she took me around and I got to see what she was doing. For a while she did fashion shows to raise money for certain nonprofits, and once she put me in there as a child model. I was immersed in this space. It wasn't organizing in the way that I think of it, but she used her skills to support people in the community. My mom doesn't consider herself an organizer because I think the way that we experience organizers is as people on the front lines or at rallies. She doesn't see herself like that. But I see her as an organizer because she finds ways to be useful for the community.

She also expects a lot and that can be good. Obviously, it gets overwhelming. But she always pushed me hard to do the best that I could in every single situation. That's why I got so heavily involved in the community, because when I expressed that I wanted to organize, she was like, either it's one hundred percent or nothing. So that's what I did.

Now, she's the executive director of Muslim Civic League, and I joke that it's like a family business because she calls me and tells me what's going on and we kind of scheme together and it's really fun. Even though we have a lot of differing views, she always has something to teach me and she checks on me a lot. Sometimes when I'm telling her something, she'll say, no, that's not how the community sees it. I'll realize, yes, you're right, I should be approaching it this way. She has a different perspective and isn't afraid to tell me that I'm wrong, and I'm not afraid to tell her when she's wrong. We've developed an interesting relationship around organizing that connects us, and we have a lot to relate to because of that.

Then there's my grandmother. I wouldn't be where I am if she hadn't gone through the struggles that she did as one of the first women to get a PhD in Pakistan at a time when women weren't graduating from high school. There's reasons why she was able to do that in the first place that come from privilege, and I have to consider where I would be if she hadn't had the opportunity to get a PhD. As soon as I started getting involved in environmental issues, she was the one sending me quotes about environmental verses in the Koran. She always watches my speeches. She calls me right afterwards. She is a pillar I rely on a lot.

At school and in my organizing, I listen to people discuss what they think a good future looks like. The more I think about it, the more I want to uplift other people, specifically Black and Indigenous people, because I feel like they are already working towards a more just future.

I want a future with not just youth leadership, but also intergenerational leadership, because I think that leads to more care for each other. This overreliance on young people for hope and solutions ignores the work that older folks have been doing for a very long time. It's also a big burden for one generation to carry, to be hopeful for everyone else.

I was having a conversation with my brother about how separated we are from the materials that we get from the earth, like when we see a wooden table, we see a table, we don't see the tree that it came from. I think that just speaks to how separated we are from the production of things that are supplied to us from the earth, and I want to see more connection with our food and materials.

I've also been thinking a lot about another reason I got involved in organizing. I think everyone feels like this at some point, that there are so many things in our life we can't control and it's not up to us. I was frustrated by that. Organizing was a way to take some sort of control. Autonomy may be the best word to describe that, being able to make choices for yourself that are in tune with the community. Choice is valuable.

Mishka Banuri is a Pakistani Muslim American who organizes at the intersection of faith, climate justice, and gender. She is the cofounder of Utah Youth for Environmental Solutions and is a current student at Middlebury College. She has been politically active since seventh grade, holding institutions and politicians accountable while empowering Muslim youth and students of color. She is a recipient of the Brower Youth Award for her work crafting and advocating for the Utah Climate Resolution, the first of its kind to pass in a conservative state.

SECTION 2
QUARTER MOON

A Quarter Moon, visible to us as a half moon, represents the midpoint between a New Moon and a Full Moon. In this section, we zoom in on key events that happened in 2020, including COVID-19, the uprisings for Black lives and prison abolition, the election, and more. These collective experiences brought brighter possibilities into focus as we began to see more clearly where we had come from and what we needed to do to build a better future.

TULSA

BY LINDA HOGAN

1.

Not the white men riots of the past,
but only yesterday
a man was shot in the back by a police officer.
She was white, he black, she a medic
who didn't help him, nor did the others who arrived.
For minutes they watched while they could have
saved him but for his skin.
I try to imagine watching a man die
because they fear or hate the darkness
of a human, a man who had no weapon, not even words,
the man who began his day like any other,
saying to his wife, Helene, I'll be back early today.
I'm taking you out for Mother's Day.
She sat under the lamp with her tea,
finishing the hem of their daughter's jeans
before she left for work. The pictures on the table
of their children, children with more and less melanin
in their skin. They are beautiful and smart, loved,
and doesn't it scare a father that they are learning to drive?
Does it scare you that one dates a white boy and they might love?
What does the officer think
as she stands, watching the man lose blood and die?
That she might get in trouble? That she won't?

2.

I am a dark woman. Dark. Darker. Even Darker.
A Chickasaw woman from the very old days,

but if the police saw me today they would think me white,
maybe whiter than them. I can pass.
They would save me, not knowing
the history in my skin
that lies to them
and how I might be thinking of them with fear
or something worse.

3.

The kids from the tribe had a chance
to go to a soccer game, so they kept up their grades;
the game was their reward. Excited, they rode the bus, so quiet,
and sat on the bleachers, learning the game, watching,
until the white men above them poured beer
on the children's color of skin, poured beer on their coats
from the unknown reservation world
from which they came.
White Men.
Native Children.
I wonder, if like the policewoman,
their soul came from some other place.

Biography on page 12.

A YEAR OF LESSONS
BUILDING THE FUTURE NOW

BY LYRICA JENSEN MALDONADO

There's stillness in the air and sky. Fires crackle and throw sparks in kitchens. People ground themselves around steaming mugs of atol sitting on wooden chairs in earth-packed homes.

I sit in a bedroom with walls textured by cement, spiders weave their homes over my head, and inside and outside blur together as the ants form a line under the door.

It's March 17, 2020. I'm in a small aldea (village) in the Department of Huehuetenango of Guatemala. My family sits in the kitchen, while I take some alone time in a bedroom, discussing what's likely on everyone's mind that night: the COVID-19 pandemic. It's been about a week since the first case was diagnosed in Guatemala, and some weeks since the pandemic caused life to be upended in the United States. I've watched news reports of the pandemic in the US, chaos and hoarding of resources, worries about failing supply chains and lack of basic resources.

Tonight all of Guatemala will shut down. The sixth COVID-19 case will stop public transportation, the airport will close down, and the government will institute mandates. Thousands of foreigners will book special flights back to their home countries in a frenzy.

I am deep in the highlands of western Guatemala, surrounded by colors and languages older than the governments of both Guatemala and the United States. My trip to visit family here is suddenly extended by a month. I am stuck in the country, with no flights back to Arizona, my home in the United States.

I grew up in the United States as a second-generation Guatemalan American. My father emigrated from Guatemala as a young adult around the time I was born. When I was in elementary school, I learned through history and popular culture that the Maya people, my people, of what is now Mexico and Central America were extinguished by Spanish and European conquistadors. I will spend a lifetime unlearning that notion as I learn the ancestral customs of our people.

It's about three weeks since pandemic hit Guatemala, and I have spent nearly two months in a small rural town close to family. The water shuts off every day. No grocery store exists within miles of my house. But I am learning so much about abundance. The abundance that flourishes when neighbors care for each other. The way Maya Mam women create networks of care through gifting fruits, seeds, cooked meals, potable water, remedies for head and tummy aches. How the land continues to provide no matter the status of Wall Street. We live in such flowing abundance when we shift our energies to our relationships with the land and ourselves. In the United States, this form of reciprocity and care is most often called mutual aid. I'm not sure if my relatives have a phrase like "mutual aid" in their language; maybe it is something like kinship. But it is as natural as the way leaves fall from the trees.

Yet, for me, it is a learning process.

After a few weeks in this small village, I realize that I too have to give back. This feels burdensome until I realize how in abundance I live. I have so much to give when I unlearn the scarcity I have been taught. I have beans to re-gift, limones, plant cuttings, mangos from the market. The Indigenous women of Guatemala have been practicing kinship and mutual aid long before COVID-19 and have continued to do so throughout the pandemic, with little change to their already existing networks. I am learning that mutual aid must not just come from crisis; we must practice kinship in times of rest and pleasure. We must build from our lives our own abundance.

In mid-May, after long conversations with family members in the United States, I purchase a flight back to Arizona. I make this journey north with a US passport and privilege. Throughout a devastating pandemic, the United States continues to deport Central Americans back to their home countries without adequate testing and safety measures. Commercial airlines, including the same airline that flies me back to Arizona, profit off of these deportations. COVID-19 cases in Guatemala are traced largely back to these initial deportations. It leaves me with so much rage that the US refuses to stop deportation proceedings, that Immigration and Customs Enforcement (ICE), the apparatus that deports migrants, cannot stop even during so much death. I do not know how to move forward from here, except for the loving abolition of carceral power structures, including ICE, prisons, and policing, which continuously design disaster for our climate and communities.

A word popularized during a summer of uprisings for justice for Black lives, abolition calls for the end—not reform—of systems that do not serve the well-being of our community. Although some see abolition as violent or destructive, abolition is a loving act.

Dismantling structures, institutions, and ideologies because they hurt people is about love. Imagining a world that isn't marred by borders, police, genocide, and white supremacy requires bravery and hope. And for many targeted by these oppressive systems, abolition is the only way forward to survive and thrive.

When I return to the US, I practice what I learned from my people in Guatemala. My time learning among the Cuchumatanes Mountains in Guatemala has restored the fight in me for climate justice in the

Southwest, a region where mutual aid networks have blossomed out of targeted communities, where youth take extractive industries by the horns and wrestle with rage and love. I take on a role coordinating a youth-led climate justice group called Uplift. This work is not easy. I struggle with other young people to connect rural youth to the internet, we regularly receive news that our family members have tested positive for COVID-19, and my own depression and anxiety spiral in and out of control as the summer heats up. While I spend time with youth from the greater Southwest, I find they also harbor lessons for me. While my relatives form care networks of material support, youth in this circle create digital moments of care and support during our daily lunch circles. Frontline youth are here, and we are not waiting for permission to care for each other.

The youth I engage in the summer of 2020 are in a fight against crises of COVID-19, racism, environmental racism, misogyny, and classism. By holding space for conflict, healing, connection, and the radical imagination of a just future, young people exemplify that breakdowns are also breakthroughs, that crises offer moments of transformation. I have heard so many slogans like these throughout the pandemic, and have become slightly resistant to them. Yet, I see our young people as vessels of creativity, imagination, and deep transformation of our systems.

I came of age during the climate crisis. I make decisions in my life based around the health of the present and future climate, but I long for careless youthfulness, for hazy nights, for laughter and recklessness. Youth are so tokenized within the environmental sector, yet rarely do we have the power to change our circumstances. So, we take the power through mutual aid, through wealth redistribution, through direct action, through the streets, through our gardens, through our living rooms and Zoom calls.

In November, the climate crisis hits my ancestral home. I watch as two hurricanes devastate Central America. The climate crisis is fueling forced displacement and creating a new class of refugees, climate refugees. Yet these are the same communities that have been historically forced from the lands through governments and private extractive industries. The climate crisis, powered by the select few of CEOs, presidents, and the wealthy, continues legacies of displacement from land. It is heartbreaking to know that our beloved lands and cultures are sacrificed for the artificial wealth of the few.

I envision a world where Indigenous people are not expelled from their lands and forcibly removed from their communities and cultures. I am so blessed to hear Maya Mam spoken by my great-grandmother and my young cousins. But as the climate crisis accelerates, as hurricanes hit Central America, more Indigenous peoples will migrate north, leaving their homelands and families. While our diasporic communities are loving and strong, we will lose those relationships to the land that bring us home. We will lose the lessons found in the deep corners of life that teach us about care and reciprocity, about resiliency and autonomy.

In our new world, we will be able to stay home. We will migrate for love and wonder and curiosity. We will follow the traditional migration patterns and embody new ones that allow us to forge strong connections into the future. Yet we will migrate with consent; mothers will not cry because their youngest son has left with no return date.

In our new world, we will bring cooked food to our neighbors, we will grow plants out of plastic gas canisters, we will decompose borders while we come home.

Our new world is here, our liberation is now. We will not wait for the future to come to us. We will build the future now in all our love, rage, and joy.

Lyrica Jensen Maldonado (she/her) is a second-generation Guatemalan immigrant with settler, immigrant, and indigenous (Maya Mam) roots. She is the co-director of Uplift, a youth-led climate justice group, and feels most responsible to the diverse landscapes of the Southwest, where she grew up, and her ancestral homelands of Guatemala. Lyrica believes that the most powerful tool we can possess in this moment is the ability to imagine an alternative future, one rooted in joy, care, and liberation.

A REFLECTION ON DISTANCE

BY JOHN TVETEN

Our computer stations are only six feet apart but I can't recall if we've ever touched. Was there an accidental bump while negotiating the ER labyrinth? Now my hand rests on your shoulder. Through double gloves, part of a body covered in synthetic armor, it is meant to be a touch of reassurance. My voice, muffled by two masks, tries to project the accompanying words, "We'll get you through this." Words I mean. Words I believe. But I can't help my thoughts from drifting towards Nieci.

Just an hour ago I was told she had died. Just a week ago it was she in this very bed. Like you, she was an integral part of our hospital. Performing a task that I came to see while I was a resident as the heart of the department: the quartermaster who plots the course and steers the ship.

Nieci was a powerful woman of color. I've heard her called "big mamma," in the most respectful light. Played out in real time she was bigger than big. Grand is more like it, singular. Brazen in her determination to provide, with arms that could hug two city blocks. It was impossible to get in her way.

While you have the same effect, your way is different. In your tradition you are the silent pillar. A cornerstone in a foundation that has absorbed the weight of generations of massacre from disease in all its forms. Providing the strength needed to nurture community and sustain culture. From six feet away, I have seen you calmly handle the next crisis that your cell phone delivers even as you steer me through the next crisis that comes through the ambulance bay doors. Now it is you who is in need of support. When I tell you that we need to admit you, are you anxious? With you it is hard to tell.

I walk the roughly sixty meters to the medical floor and I reflect on distance. Six feet, the distance from middle fingertip to middle fingertip of the average English sailor's outstretched arms, aka a fathom. Six feet, the distance declared to bury victims during the plague of 1665 to prevent further spread, hence the euphemism 'six feet under.' Six feet, the standard unit of social distance…based on what? A sneeze can propel droplets to speeds of one hundred miles per hour. According to an MIT study, these droplets can travel eight meters and remain suspended in the air for ten minutes. A cough is not far behind traveling up to six meters. Why six feet? The answer is no different in 2020 than it was 355 years previously; it seems about right.

Whether I measure by seven sneezes, ten coughs, or thirty-three social distance units I arrive at the COVID ward. I walk past the void left where Nieci used to sit. It is fathoms deep. I can't see where it ends. Three days have passed since I last saw you and a quick glance tells me your oxygen requirement has gone up. Your anxiety is now plain to see. Can you see mine?

This disease is like nothing I've encountered in my twenty-five years of practice. For most it goes unnoticed or is a quickly passing storm. But some get walked to the canyon rim, are allowed a moment to take in the view and then without warning are cast into free-fall. I wish I could tell you that someone couldn't be you.

Double-masked and gloved, armor in place, I sit on your bed and hold your hand. It is the best medicine I can offer. There is no distance. I remain hopeful. You remain strong. We make a pact to hug.

John Tveten is an emergency medicine physician based in Flagstaff, Arizona. He began his career working in Emergency Health on the Hopi and Navajo Indian Reservations in Arizona, where he began to explore various art forms as a way to process the traumas that flow through the ER doors. He has been a photographer and filmmaker on multiple continents and currently serves as the executive director of the Flagstaff Mountain Film Festival. He is a musician and plays in two regional bands. Writing has been the art form that has enabled him to find the most perspective and ultimately empowerment from his work in the ER. He has just recently begun to submit his work for publication.

WHAT HAPPENS AFTER THE PROTESTS

CONVERSATIONS ON ANTI-BLACKNESS

BY FRANQUE BAINS

As told to Alastair Lee Bitsóí and Brooke Larsen
January 2021

When George Floyd died, I felt like everybody, with COVID-19 happening, had more time on their hands to be aware of anti-Blackness. People aren't running to the bar or going to their soccer practice. They're not doing their regular schedule. People had more time to be aware. Because anti-Blackness happens all the time. Months before George Floyd was killed, a young Black man jogging in Georgia was shot because some neighbors thought that he was trying to rob homes. He was literally just jogging. That happened months before Floyd's death, but the uprising didn't happen then. The same week George Floyd was murdered, there was a Black gentleman in Central Park looking at birds, and this white woman outrageously makes accusations and leverages her power in a really ugly way—she threatens to call the police, and she does call the police and makes false accusations out of her own fragility. This stuff happens all the time.

In 2020, people slowed down enough to see that this shit sucks, and it catapulted a national uprising when George Floyd died in an atrocious, violent, and awful way. We are killed at much higher rates—that's the whole deal with Black Lives Matter. Women die giving birth at higher rates in the Black community than any other demographic, and

Black children have much more struggles, whether it's suspension or academic performance in schools. So we ask, what's going on? That's the premise behind Black Lives Matter. It's not about all lives matter because that's inherently a part of the human experience. All of our lives matter. All marginalized people matter, all people matter, period. But, Black lives are not treated as if they matter in America. We have data and statistics. These disparities are glaring and true and can be measured, and they shouldn't exist. That's the intent behind the messaging.

I have a lot of experience teaching and holding conversations, so in the summer of 2020, I thought, what can I offer up to the community? I felt conversations would be a way to answer the question, what happens after the protests? I started the Utahn Conversations on Anti-Blackness Project because we've been doing this, asking this, for hundreds of years. It started getting sexy in the abolition movement in the mid-1800s with the Underground Railroad. People began to actively address misinformation about who Black people are and what they're capable of and their place in our society, because there was a strong misinformation campaign that fueled racism and slavery. We've been fighting this for a very long time, the lies about the Black community's abilities and African people's role in the community. For centuries, pseudoscientific ideas about Black racial inferiority have persisted to justify the cruelty of slavery or discrimination: our brains are shaped differently, so we're not as intelligent, or even *Black people like being enslaved.* The religious implications were, oh, they're savages and they need to be saved. Totally unfounded layers and layers of anecdotal evidence that started in the fifteen, sixteen, seventeen hundreds justified the atrocities of slavery. But, people began to say, no, this stuff is not true. We've been working at this justice campaign for hundreds of years now.

BIPOC people still experience terrible bias in the workplace, all the disparities are there. Doctors don't listen to Black people, and then don't know why we don't want to go to the doctor. We are treated awfully. How do we change hurtful behaviors? We have all of this work we've been doing for decades and decades and decades, but our behaviors are not very different. We're doing a lot of justifying, a lot of gaslighting, a lot of fragile bullshit that blocks the progress we need so that people experience actual changes. For example, a kid should be in a classroom where they're loved by their teacher. A huge criticism of integration is that now our children don't have teachers who love them. When we were segregated, we had way less resources, but we had adults in the room who respected, saw us, and loved us. Then in the '50s and '60s, when people were integrating the schools to have better access to education, Black students were being spit at, yelled at, denigrated. It was absolutely atrocious. Today many things still function the same. Black people are in jails at higher numbers, die at higher numbers, and we aren't seen. How do we change that? We're doing a lot of work, and we don't need to waste our time anymore. How do we actually make these changes? We know it's a slow process, but when you name it and acknowledge it, change can happen.

It's challenging, though, because when you begin to do the work and you begin to have the conversations, you, as a person of color, begin to get a lot of pushback. People get so fragile and there's just no trust, but we have to trust the mouths of our BIPOC community because they're experiencing it front on. Some people think that things are better because we have laws in place, that we don't need affirmative action to correct for historical exclusion and discrimination. An affirmative action proposition didn't pass in California in 2020. That was shocking for me because California is supposed to be such a liberal, progressive place. The measure would have removed the state's ban on race and gender considerations in public hiring and college admissions. I think you have to be misinformed to not support policies that balance and equalize the system. I've seen many people pushed out of workplaces because of their values and because they don't have support to voice their opinions. It's related to the violence that BIPOC people experience in our workplaces and our schools. We can look at the data and say, look at the disparities, what's happening with our Black community? The desire in Utahn Conversations is to hear this story. Even in nonprofit organizations that say they're addressing racism and have their DEI statement, people are often not willing or able to talk specifically about anti-Blackness.

Utahn Convos was offered up in June, just a few weeks after George Floyd's death, to respond in a timely manner and give the community a way to deal with our discomfort. I don't think I've been in a space that does something so specific. We required two things: participants need to be in Utah and agree to discuss anti-Blackness. It was helpful to be specific in that way.

We implemented a distributed model where trainers coached a larger group of conversation hosts who then each recruited ten to twelve people to participate in a series of three conversations. We also asked people to identify and convene in affinity groups based on race and other shared experiences. You can't underestimate the damage that white folks don't mean to do when they are talking about race in a mixed-race group. People have a lot of work to do, and they often put it on Black or brown people. Especially in someone's early stages of their journey, it's really best to have people discuss anti-Blackness in their own affinity groups, to keep everybody safe. By focusing on anti-Blackness in safe spaces, we got to get real about what Black people experience. That lifts up all communities. We had wonderful leadership from the LGBTQ community. We had people from the Mormon community, people from the liberal community, people from the conservative community, and each person who was holding conversations had a trainer to help them walk along the way. Because we know this is going to be hard and it's a community thing.

I wanted these conversations to help people understand themselves, as well as anti-Blackness in our society, so that behaviors can really change. At our first meeting for Utahn Conversations, we realized people want to talk to their cousin who is right wing, and hash it out and figure out how to get their cousin to understand. They wanted to talk to their dad who supports Trump and really convince him, to come with the facts so they can change his mind. But that's not what this process can do. We called an impromptu meeting to clarify that these conversations are offering up a community of care to talk about anti-Blackness with people you trust. It's about growing together, about moving away from this position of "I know something and I need to teach my dad," towards acknowledging that "I wrestle with this too." Because you do! If you didn't, then our workplaces would look different. Listening to my BIPOC friends, I know there are many nonprofits that are purported to do such good work for our communities. They've got brilliant mission statements, but systemically, they're not able to show up in so many ways. No matter your workplace or community, Utahn Conversations was about you and your own accountability and being with people to have that courageous conversation.

How do I show up? In the body. Resmaa Menakem, who wrote *My Grandmother's Hands*, had this amazing podcast on *On Being* that came out like a day or so after our first training. He explains how what's in our body influences our behaviors, so we get grounded in our body and listen to our body and feel our feelings and do that work so that our behaviors can change.

Organizations need to do this work, too. For instance, nonprofits seem to chase dollars in ways that make them so busy. So you hear, "We got a grant. We're doing something now, but we have a million other things on the burners, so we're going to tend to this, and we'll get back to you." I don't know why nonprofits have to do so much. I don't know why they have to be moving so fast that they burn out their employees. Often the leadership doesn't even have the capacity to address issues that arise. By the time they do, it's long past the incident. It's been festering and boiling. You don't know what to do. You get a grant to do DEI work, and then six months later, you have a person come in to help. But what have your BIPOC employees been experiencing from the moment the stuff started to happen to six months later when the person comes in to have that first meeting? Sometimes I wonder, why don't the nonprofits just turn off the burners? Turn off the burners and say, "Man, this one needs my attention right now. Because I care about this person." I think they'd be surprised at what happens when you do that.

When you see someone hurting, stop. It's a big practice of mine that I'm trying to do at the small scale. I was at the post office and this postmaster was being so rude to this older lady. I thought to myself, I got to get some work done, I got to get this shit mailed out. So I didn't say anything. But, did I have to get my shit mailed out that day? I could have just stopped and had the conversation that I wanted to have and say, "Hey, you're being really rude to her," and be prepared for whatever repercussions happen. She didn't need to be treated like that, and I had the opportunity to just stop.

It's hard to slow down, to breathe. Utahn Conversations was a big lift for people, so how do we have the bandwidth to keep it going? How do we keep asking people to do such hard things and make it sustainable? How do we help people have the capacity to keep this work going? How do we get people to understand some of the anti-Blackness values? It was hard to teach people that behaviors can change if people learn how the body experiences feelings. How do we make all that accessible for people?

It's a scary time with the insurrection at the Capitol, the harmful behaviors and actions of someone who was elected president, and the confusion and the division that make people believe a plethora of misinformation that is permitted to be posted on our news outlets and on our social media. Solid misinformation. And it's so dangerous right now. As writers and creators, we get to envision this world and what it can be, so we can speak truth to power and get busy making the future we want to see.

Franque Bains is a poet, an organizer, a google doc nerd, and that gluten-free friend. She is convinced that the key to happiness is bringing your ideas to life and helping others do the same. She currently works to build community through storytelling in Salt Lake City.

LATINX POLITICAL POWER

A YOUNG ORGANIZER'S JOURNEY

BY IRENE FRANCO RUBIO

When I was in the second grade, my best friend's dad, an undocumented immigrant, was picked up by ICE as we headed home after a long day of school. Ironically, I recall learning about the Founding Fathers that day, yet the American values we were taught and the Pledge of Allegiance we were told to recite in class every morning didn't seem to align with the so-called democracy we experienced in our lives. This was in 2007, when Donald Trump was still playing a successful businessman on TV. I grew up in Arizona, where we were living in Trump's America long before he was ever in office.

I'm a twenty-year-old woman of color of Guatemalan and Mexican descent, and I grew up in Joe Arpaio's Arizona. Under the former "Sheriff of the Southwest," state-sanctioned violence, fear-mongering policies, separation of families, mass deportations, and voter suppression were the norm. Nine years after my friend's dad was taken, when my community heard the news of Trump's nomination, we knew what America was facing. We knew that if he could, Trump would turn America more into Arizona. However, we also knew that people-power organizing could remove hate from office. After serving twenty-four years as the thirty-sixth sheriff of Maricopa County, the fourth largest in the country, Arpaio was voted out of office in 2016.

Though it often goes unnoticed in the national political conversation, Arizona has a voter suppression problem. Before the Supreme Court struck parts of the Voting Rights Act in 2013, Arizona was among the states and localities with a history of race-based voter suppression that were required to clear any changes to voting laws with the Justice Department before they could take effect. At the start of 2020, I joined other local voting rights organizers at the Arizona State Capitol to sit in during Senate Judiciary Committee hearings multiple times a week. Under a Republican majority that was known for racist policies, our presence was vital to defend and protect our communities. Proposed voter suppression bills would have prevented our community from exercising their right to vote in the 2020 election through fear-based and voter intimidation tactics. Legislators even introduced a version of Senate Bill 1070, Arpaio's notorious racial profiling and the nation's strictest anti-illegal immigration law that targeted predominately Latinx communities across the state.

Soon after, I became more involved in the fight to protect voting rights and sought to engage diverse communities of color in the electoral process to overcome the belief instilled into oppressed communities that voting does not work. I was an organizer for When We All Vote, Michelle Obama's non-partisan voter registration initiative, and I worked directly with young high school students in school communities across the country. In a very Gen Z student-led type of way, we recruited and registered over thirty-five thousand seventeen- and eighteen-year-olds, contributing to the historic surge for turnout among young people and people of color. My passion for community organizing and investing in young people grew even more, and I was able to recognize first-hand just how great of an impact we can have when we as BIPOC communities recognize the unique and individual powers our votes and our voices hold.

Due to the monumental efforts of community organizers to get out the vote, Arizona turned blue. The Biden and Harris ticket won, largely thanks to young people and Latinx, Black, and Indigenous women who mobilized their communities. By organizing community initiatives, mass movement building, strategic campaigns, and simply canvassing historically marginalized neighborhoods, BIPOC organizations set a new standard for what's possible when we organize and uplift progressive values in states viewed as conservative on electoral maps. The wins at the ballot box reflect a movement that took years in the making. For example, the important wins for Democrats in Georgia were built on a decade of organizing and strategizing by people like Stacey Abrams. Long before I ever knew anything about the Republican political landscape where I grew up, community members across Phoenix were realizing and utilizing their political power without knowing the monumental impact their power would have in the years that followed.

When I was growing up, my parents, avid news watchers, provided me with critical exposure to the world via local antenna TV channels, without sugarcoating reality. I'd hear my dad talk about politics and conclude there was a correlation between the issues being discussed and the color of our skin. He tended to become more expressive about topics involving People of Color, whether the coverage portrayed us accurately or stereotypically. My mom spoke with hope, knowing that people were hurting within and beyond our community. Together they taught me to be aware, useful, and they inspired me to be part of a movement that was absolutely needed. Despite the political climate and racism we encountered in Phoenix, they ensured me that we as a Latinx community would prevail.

I knew the Latinx population across the US consisted of incredibly hopeful, motivated, hard-working people. I recognized my dad's work ethic and his hustle as a truck driver, traveling to and from Los Angeles every other day in the blue semi-trailer truck in which he'd invested so much money and time. I recognized my mom's hustle in working tirelessly to raise my siblings and me, in addition to her job at a gas station company warehouse.

My parents faced circumstances that were familiar to many working-class families across the West Valley in Phoenix. It felt as though we had limited access to opportunities for something grander— if those opportunities existed at all. I came to understand that this was an injustice, that we needed to overcome this oppression, and that my community was already making strides to fight it.

As a teenager, I became heavily involved within my Phoenix community while reporting and writing to uplift diverse voices as an aspiring journalist, fighting against voter suppression bills on a weekly basis at the state legislature, advocating for voting rights, organizing for social and racial justice, and having critical conversations about political activism and civic engagement. Those experiences allowed me to gain a solid understanding of the importance of being civically engaged, with a particular focus on voting and educating others about how to make their voices heard, too.

Even though my parents first exposed me to political conversations, I wasn't raised in a family that believed in the power of our votes. Being the forever curious and relentless child I was, I began to ask questions about voting and politics, despite the pushback I encountered. The stigma around voting is not exclusive to one particular community, but it is rather common across various disenfranchised communities. Many Latinx and similarly historically oppressed populations have low voter turn-out, often because they had been hopeful about change that was possible only to be disappointed. Even when Latinx communities vote for progressive candidates, such as Barack Obama—initially a symbol and beacon of hope for BIPOC communities—we are often left with broken promises and failed opportunities for advancement of our communities. For Latinx communities in particular, the immigration infrastructure we were witnessing in Trump's America, the detention centers, had originated during the Obama administration, a testament to why folks in Latinx communities felt let down and chose to not engage in the political process.

I was raised in the ghettos of Phoenix's West Valley, within a predominantly Latinx community that was influenced by socioeconomic inequalities and held a general distrust toward the government. I understand the complex dilemmas and beliefs people of color and systematically oppressed communities continue to face because I have lived through them. But I also know that voting gives us the potential to elect local, state, and federal level officials who reflect our values and keep our best interests in mind. These elected officials serve on various committees that make vital decisions about our communities. If they are held accountable, they vote for and enact policies on behalf of their constituents based on what they believe is best for the community they serve. If we want to see our community's values reflected by those in power, we must show up at the polls for every election, not just the presidential ones.

In the aftermath of Trump's nomination, my parents were much more open to the idea of voting despite their general distrust towards the government. My involvement in the movement motivated and educated my parents to become further engaged, but that's not every family's reality. These potential voters must be reached, and the organizing must continue beyond presidential elections. Elected officials must also invest their time directly into getting to know our diverse communities and not just pretend to care about their concerns when elections roll around and it's suddenly convenient. By authentically earning trust and following through on promises, elected officials can have the opportunity to generate long-term support and sustained civic engagement. BIPOC communities typically don't engage in electoral politics, not only out of fear of being betrayed, but also because of continued distrust towards the government due to generations of failed promises from politicians. It's time to rebuild relationships between communities of color, the government, and the political process overall and establish a new era where the democracy we engage in is actually by and for the people who inhabit it.

Voting may seem inadequate or not effective in seeing immediate change, but voting is merely one aspect of creating the type of change we hope to see in our country.

Even though social justice movements defeated Trump in 2020, we still face the white supremacy Trump emboldened. The insurrection at the US Capitol that occurred on January 6th, 2021, showed what's possible when you're a white supremacist in America: storming a government building in order to harm the nation's leaders with the support of the outgoing president who incites a riot with lies.

I called my dad as I watched it happen, knowing he was watching it too because of his continued concern in American politics and hopes for a more democratic future. A part of me was in disbelief at the sight of white American men destroying their very own federal government's property in a country that they supposedly "loved" and would "die fighting for."

We were amazed at the lack of security response and law enforcement's inability to fight back to protect the nation's most sacred building, a fortress of our democracy. Above all else, we were shocked to see that this was a battle of the same oppressive forces in combat against one another. White men leading an insurrection and upholding white supremacist ideologies versus white men in positions of power commanding law enforcement officials to back down demonstrated a fascist system that solely prioritizes and upholds white Americans, not the entire body of the people it governs.

In 2019, I had the unique privilege of interning on Capitol Hill for the US Representative Deb Haaland, now the first Indigenous Secretary of the Interior. I walked the halls of Congress as a low-income, underserved youth and first-generation American while working at the office of one of the first Native American women to serve in the House of

Representatives. It was a historic experience, being a young woman of color learning from and looking up to another woman of color in an elite space dominated by white men who intended to keep women, and women of color especially, out of positions of power. This opportunity gave me the ability to recognize my unique power as a young Latinx woman to overcome any and every challenge to create more access to spaces of privilege, like the US Capitol, for BIPOC folks like myself.

During the first impeachment of Trump, I invited my parents to visit DC, to share with them my experience of walking the halls of Congress every day. I gave them a personal tour of the US Capitol. We began by the front steps outside and eventually made our way to the center of the dome, taking a break in the Rotunda to admire the breathtaking view of the painting floating above us. My dad said, "A venir a los Estados Unidos como inmigrante, nunca me hubiera imaginado estar afuera de este edificio. Pero explorar el interior y caminar por estos pasillos históricos, nunca me hubiera imaginado." As an immigrant, he would never have imagined walking these historic corridors.

They admired the building and were thrilled to have the opportunity to be there. We walked through the pathways I had memorized over those past few months till we stood at the heart of the building. Through their journey as undocumented immigrants turned American citizens, and now visitors in the people's house that they could rightfully call their own, we recognized the privileges we had been afforded and felt especially grateful in that moment for the "American dream" they had come so far to obtain.

When I watched the white supremacists storm the Capitol, I thought of that day I gave my parents a tour of the same building. The insurrectionists had the privilege to cause destruction without facing any severe repercussions for the physical and societal damage they caused while people like my parents in the Latinx community and those in BIPOC communities will continue to be treated as second-class. They are more likely to be killed directly by the nation's forces than protected by them. Regardless of the significant contributions Latinxs continue to make to American society, we will not be prioritized until our values and communities are reflected in the very systems that govern us.

There's so much hope and possibility for the future, but it is essential for the people to be uplifted, heard, and given more than a few seats at the table. Those seats should represent the population served so the interests of BIPOC inform the solutions. Representation matters, but we also must go beyond that. It's critical to make it into existing spaces of privilege, prestige, and power that are not typically inhabited or designed by BIPOC folks so that we can dismantle the existing systems and create new ones authentically designed by us and for us that work for everyone.

While women of color should be given the power to build new systems, they should not have to carry the burden of ending white supremacy. In the 2020 election, with the highest voter turnout ever recorded in American history and with the most diverse coalition of Democratic voters, we still barely managed to win the presidency and the Senate. Through a national collective effort, we fought and saved our democracy, but it nearly slipped away from us and we must ensure such a close call doesn't happen again. Fifty-eight percent of white voters voted for Trump. White people need to talk about whiteness with one another, take responsibility for their contributory or complicit behavior, end the violence of white supremacy, and surrender their power.

In 2020, we witnessed the effects of climate crisis through the ongoing wildfires that consumed the Western US for months, deadly force by local law enforcement towards Black communities, and inadequate leadership reluctantly provided during a public health crisis. America was finally forced to bear witness to the side effects of the society it inevitably created. Quite frankly, the US was long overdue for a reality check.

Activism became trendy in 2020, but to truly dismantle systems of oppression and build a more just and caring future, everyone must continue to remain involved, vigilant, aware, and socially conscious. When people choose to not engage in our community, the democratic process, preserving our environment, dismantling systemic racism, and demanding more than representation, they become cogs in the industrialized American-made machine.

It's sometimes easier to disengage, so we also need to care for ourselves and one another to make social change sustainable. I've learned to work constantly and often more than I can handle—a generational practice and habit passed down to me from my parents, a work ethic that is pretty much ingrained in the way Latinx communities function and exist. I wholeheartedly believed in working myself down to the core because I assumed there was no other way to survive in a society that was not designed for someone like me to thrive. Soon enough, I came to understand how my community organizing and activism are vital to the advancement of not just my family or ancestors, but for every BIPOC community in this movement for change. But I now know that I must take time to reflect and rest in order to continue to serve the communities that have worked nonstop to lift me up for generations.

During this pandemic, I've learned a few things to uplift our historically oppressed populations and sustain our long-term organizing and activism efforts. The first, is to take time for oneself. Take time to think, slow down, and be present in the moment; overworking will cause more harm than help. We don't always need to produce, we just need to be present, exist, and merely survive in an American society that is preoccupied with trying to suppress us. Against all odds, we as historically oppressed people have continued to survive, be resilient, and approach our work every time stronger than before—and that is worth celebrating.

Irene Franco Rubio is a social, racial, and environmental justice activist born and raised in Phoenix, Arizona. She advocates for BIPOC communities through her writing, digital community organizing, and intersectional movement building. Irene is a young woman of color of Guatemalan and Mexican descent devoted to the movement for progressive social change.

CONFRONTING INJUSTICES IN NAVAJO COMMUNITIES

FROM URANIUM MINING TO VOTING RIGHTS

BY TARA BENALLY

As told to Alastair Lee Bitsóí and Brooke Larsen
January 2021

In January 2021, the same day white supremacists attempted a coup at the US Capitol Building, Tara Benally talked with us about her fight to end voter suppression in San Juan County, Utah.

San Juan County, Utah, is the size of New Jersey, but fewer than fifteen thousand people live there. The population of the southern portion of the county is almost entirely Diné and Ute, home to both the White Mesa Ute Mountain Ute community and the northern reach of the Navajo Nation. The northern part of the county is mostly white and has been the site of anti-government protests by armed white people.

In 2016, a US district court ruled that San Juan County violated the 1965 Voting Rights Act and the US Constitution through racial gerrymandering. The county had relied on race to draw the boundaries of voting districts, systemically weakening the Native vote and preventing Native people from holding power on the county commission or school board. Leonard Gorman, the executive director for the Navajo Nation Human Rights Commission, sued to redraw the county commission and school board districts. He won both suits. After the voting districts were redrawn, two Diné men won seats on the county commission in 2018. Thanks to Tara's efforts to register voters and get out the vote, San Juan County now has a Native-majority commission for the first time.

I'm of Hopi descent, born for Navajo. My grandmother about five generations back was taken from Hopi and brought to Navajo. She was brought to be a slave, but the Navajo lady that took her in decided to raise her as her own. That's how we came to be here in the Navajo Nation. My great-great-great-grandmother participated in the Navajo Long Walk down to Fort Defiance. She was born over here, just north of the Bears Ears, so we have ancestral ties to Hopi and Navajo as well. That's how I would like people to acknowledge me and understand where I come from, why it's important to me the work that I do here on Navajo. I hope to bring in my Hopi ties, and my kids are Blackfeet and Paiute as well. For me, including smaller Indigenous nations in the work that I do has been important. I want to show my kids the opportunities to carry on the work that I have done, should they choose to in their future, and I want them to know where they come from.

When I graduated from high school in the early '90s I wanted to see who I could be and find that person. I knew at that time I was capable of much more than what I was perceived to be. I grew up both off and on the reservation. My siblings and I lived off the reservation because my mom was going to school at the University of Utah or down in Phoenix, and my dad worked as a steelworker and had to travel a lot for his work. When we were on the reservation, my grandmother instilled in us, "You just don't walk by somebody who is in need of your help, you stop to ask what is wrong." You have to have humility, compassion, and you have to have consideration for that person's background and understand where they're coming from. You don't have any idea what their story is and you take the time to find out.

At the time, families were receiving material for their homes to be built. A lot of the materials were being stolen, or they were weathering out in the open and going to waste. On the day that I was leaving for college, I told myself, I want to come home and help finish the homes, which are vital for our families. There was absolutely no need for wasting material and resources when there are people living in the conditions that they were living in—dilapidated shacks, multiple families living under one roof.

After being at college for about a year and studying civil engineering, I began to get impatient. One day I decided I couldn't do this. I can't sit here and have my families and my community members wait on me for four more years. I decided to quit and signed up with the Job Corps and received my certificate of completion in carpentry. After working several years in various construction trades, I went home to the Navajo Nation.

Returning home was the greatest decision I had ever made. Of course, life tends to take you down this road that will detour and take you down different paths. After several years of being home, I didn't get the opportunity to work for my community until 2013 when the Utah Navajo Trust Fund hired me. Getting that position was in itself a struggle. Every time there was a job opening, I was denied. There was always some other guy that was hired for the job or the excuse that I was overqualified. On the third attempt to hire on with the Trust Fund I got on the phone and asked why I hadn't been considered for the position. I wanted to know why I was being passed up for a job I knew I was capable of doing. After a while I was asked to do an interview along with another lady who was coming in that morning for the same job I had applied for. I wasn't going to pass up this opportunity, so I agreed. A couple of hours after the interview, the administrator called. He wanted to hire me as a construction supervisor to run a couple of crews and help to finish up some homes. I wasn't looking for a supervisor position, I wanted to be on a crew and actually do the work, but all I knew at that time was I wanted to work for the Trust Fund and start doing what I had promised myself and the people I would do all those years ago when I left for college.

Over three years, I helped to complete thirty-three homes across southern San Juan County, throughout the seven Navajo chapters in Utah, and I built a rapport with my community members. They were familiar with my work and knew I was capable of completing a project in a timely manner. Out there, people know who I am and where I'm coming from, which is very important when working with the people. You have to gain trust and respect. I made it a point to listen and understand what their day-to-day life was like. Building homes at that time was important, but I soon realized there was so much more to do. Quality education was a concern; there were also concerns about having to travel six hours to see a specialist, so we needed road maintenance. The list just got longer and longer.

When I left the Utah Navajo Trust Fund in 2016, my mom had just become a board member for Utah Diné Bikeyah, a nonprofit organization. I was taking her to one of their meetings in White Mesa, and I noticed a newspaper from New Mexico saying that one of the uranium sites there had received Superfund status for the cleanup. After I finished reading that article, it made me curious about the sites here on the Utah portion of the Navajo Nation. As I started researching, I realized there were over five hundred uranium sites on Navajo alone. More recently, I learned there are a lot more uranium sites than that, over one thousand uranium sites on Navajo, but because some are so close together, they get counted as one. That was a big eye-opener.

The first time that I had any type of experience with uranium, or the side effects of it, was when my aunt, who was two months older than me and had Down syndrome, passed from leukemia. When she was first diagnosed with leukemia, I was going to school in Price, Utah. Being Down syndrome, Navajo, and raised in the Navajo Nation, she didn't have a whole lot of services where health care was concerned. Instead of taking her to a treatment center up in Salt Lake City, she was just placed in the senior citizen Four Corners Care Center in Blanding. That's how she lived out the rest of her life, in a facility with nearly no medical attention and with doctors who didn't really know what they were doing for her. Seeing her go through so much pain and suffering, absolutely no need for it, broke my heart, knowing there are facilities worldwide where people with cancer were being taken care of and they were surviving.

My brother's older sister had been diagnosed with cancer as well. She had passed. I had other family members on my mom's side, her cousins, who were also diagnosed with cancer and they all passed. On my father's side of the family, I have relatives who were diagnosed with pancreatic cancer and didn't survive.

When I picked up that article in 2016, I started to ask a lot of questions regarding uranium. I met with Ida Yellowman, a Diné nurse, and she told me about how she was a caretaker for a lot of these cancer patients who were exposed to uranium while working in the uranium mines. Anytime uranium and Native Americans was the topic it brought tears of pain that welled up in her eyes. The pain is real. Our people are going through so much pain and suffering on a daily basis. Ida would like to see something done for them to ease their discomfort. She talks about trying to find a way to elevate their voices and make something happen. Individuals like Ida inspired me to research uranium, the cancers, and the clean up of the hundreds of toxic sites across the Navajo Nation.

My research slowed and after a while stopped in early 2018 when the Rural Utah Project approached me and asked me to work for them to do voter registration. Uranium is still something I believe I will come back to. Anytime uranium comes up in a conversation, I make it a point to sit and listen. It's still very important to me, and I want to hear our people's stories. One day, I'd like to put all of their parallel testimonies in a book, or an article, and help to advocate for our loved ones who are currently still suffering from Western society's greed.

When Rural Utah Project first approached me, I literally turned them down about four or five times. I said, "No, this is not my area of expertise. I have absolutely no business getting political." But as time went on, I thought about it. I had worked a couple of years before with Leonard Gorman of the Navajo Nation Human Rights Commission. We worked together on a voter survey across four states. The feedback showed how and why voters weren't casting their ballots. My idea of voting was just be there on Election Day, cast your ballot, go home, and call it a day. But as I sat and listened to what Rural Utah Project was planning on doing across southern San Juan County, and Leonard Gorman, at that time, was doing his presentation on voting rights issues and suppression here at the local chapters, I saw how their work tied in with everything that I was hearing when I worked for the Utah Navajo Trust Fund and talked to my community members. All of that was covered in voter registration. The bottom line was that voter registration was the way to make it possible for my community members, my people, my families to be that much further ahead.

We had the opportunity to elect two Navajos—Mark Maryboy and Willie Grayeyes—to the San Juan County Commission. When I heard Mark Maryboy and Willie Grayeyes speak, I thought to myself, this is exactly where I need to be, this is where I need to go, this is the start of something big. I didn't know that it would come to the point of being in *TIME* magazine's person of the year issue just a few years later.

When I started working for the Rural Utah Project, I had to collect one hundred voter registration forms that week. That was the goal. I planned on going over and above the one hundred voter registration forms. We started doing a lot of voter registration at the clinics. People stopped and were curious: "Why? Why do you want us to register to vote? Why?" There's so much skepticism out there. People feel that no one has helped. The federal government hasn't helped, the tribes haven't helped. Why? They might have registered before and gone to vote, only to be told they're not registered. As I started talking to the people and hearing their concerns and why they didn't vote, I felt like I needed an answer for them. Why was the work important? At the time, I didn't feel it was important. I was just doing it for a paycheck.

There were so many reasons why people didn't want to register to vote. But through educating myself about the whole process, I was able to have answers for the community members that had asked me, "Why?" I could then give them the knowledge and the tools that they needed, so when Election Day came around, they were able to stand their ground.

Also, when Willie Grayeyes and Kenneth Maryboy decided they were going to run for district two and district three, that gave me a platform to stand on to show the people we have the capability of putting two Navajos at the commission level into office. Shouldn't we have somebody who knows what we're about, where we're coming from, what our setbacks are, speaking for us? Not some non-Native in office who doesn't come around to see us or participate in our meetings. That was something that I was able to stand on, which I felt opened the minds of the community members here in San Juan County, Utah, and helped to make the people realize that voting is an opportunity for change.

Change is something that they had asked for, for years, even decades. We were able to elect two Navajos into the county commission office here locally in San Juan County despite barriers and opposition. Being in construction, which is predominantly male, I learned early on that I had to choose my battles. I had to learn how to argue what I needed to argue, make my point where I needed to. I still utilize that tool, to argue where I need to and leave things alone where I need to.

We still hear that people want change. They hunger for more because they haven't had it for so long. And now they're just barely starting to see that happen. But it's happening. Yes, it's slow. But it's happening with the community members' help, as we saw in Arizona, where we were able to flip the state to Biden in 2020. I'm very appreciative of the organizations and the people that I have worked with who have supported us through the last couple of years.

Change is possible if you just take the time to reach out to people and sit down and listen to their story and understand where they're coming from, and in turn tell your story and where you're coming from. In Navajo, your story has to come full circle first before you get down to business. A lot of the times, when we were doing voter registration, it took an hour to sit down and talk to a single individual to get them registered to vote because they wanted to hear where we were coming from and why we were doing the work that we were doing. We were able to just sit there and understand and be open-minded and then have the answers to help them see why it was important to be a part of this process.

People would say, "Oh, you're just a bunch of local Navjos coming in on our part of the reservation, and you're having the white guy tell you what to do. You're being bought just as well as everybody else that's in leadership." But we were able to change that whole thought process. We were able to be out there every day and not give up. When we had doors slammed in our faces, we were able to take time to come back and say, "I'm back again, because this is really important. This is why we need you to be a part of this." And people listened.

I think the biggest factor why people were so eager to listen to me is that Navajo lady that took my grandmother in those five generations ago. I introduced myself as being thankful for that Navajo lady who didn't have to take my grandmother in all those generations ago and raise her as her own. I'm thankful for it. If it wasn't for that grandmother, and that Navajo lady, where would I be? Where would we be? It might have ended with her there. But to be alive today because of that Navajo lady made me want to say thank you to Navajo for allowing us to be here, part of Navajo. That was more or less out of respect and to say thank you because when I knocked on doors and introduced myself as Hopi, a lot of the community members in southern San Juan County were like, "Hopi, what are you doing here? We don't get along. Why are you here?" So I would tell my story, and they would really appreciate that. People responded to that. That's what we need today, to appreciate one another and understand where each of us is coming from.

Getting out the vote was a lot harder in 2020. When the pandemic hit, the whole idea of being shut down didn't sit well with me. The one thing that I knew I had to continue doing was the voter registration, to get my field organizers out there and continue voter registration. I had to go back to the drawing board and figure it out. I had initially just decided to hire two, maybe three people in February, but by the end of February, I felt like I needed to hire everybody, all twelve of the field organizers that I had planned to hire for the year. I had planned to hire three in February, three in March, and so on until I had twelve field organizers. By the end of February, COVID was knocking on our door, and we thought we were going to get shut down. I had to fast-track everything and hire people, get them on board, and get to work. I only had a week to train my field organizers.

Part of our organization helped out with mutual aid; they helped to deliver food boxes and supplies. I delivered food boxes, but I also wanted to make sure the people knew that I wasn't giving up on them. Once I started doing voter registration, it became my one mission in life to elevate my community members' voices, not just here in San Juan County, but in Arizona because I had seen the living conditions of my neighbors in Arizona. Their stories gave me that much more reason to continue the work that we were doing despite the pandemic. Even though we were splitting our work with the mutual aid, there was still going to be somebody who would come around and say, "I need your voter registration. I need you to help me so I can help you guys elevate your voices, because this is a very important race."

At the beginning of the pandemic, people were hearing in the news how Trump was not taking action to protect the people where COVID was concerned, so when we were able to go back out in the field after the first shutdown, people started to really respond to us. We reached out to the people we registered prior to the pandemic and asked those individuals if they could reach out to their friends and family to see if any of them still needed to be registered and if they were okay with sharing their contact information so we could reach out to them. People called us and said, yes, they would like to register to vote, because they were starting to see what the Trump administration was doing during the pandemic.

We put a plan together to get gallon-sized Ziploc bags that would include a voter registration form, pens, and our field organizers' contact information. We would call ahead, let them know that we're dropping off a Ziploc bag with the voter registration forms, and we'll hang it on the doorknob. Once they filled it out, they would then call the field organizer in their community, and we'd go by and pick up the voter registration forms. If there were any questions, they would reach out or call the field organizer. That was the process that we started.

Then the second shutdown hit, and we were like, you know, great. We were just twiddling our thumbs. But by then, my field organizers were wanting to make this happen, to keep up the momentum, because before the shutdown, we registered one thousand people. And they're like, "We don't want to stop. We want to continue. We want to be out there. We know the importance of this and why it needs to happen." They had heard the stories of what we had done back in 2018 with the San Juan County Commission, so they were excited to do voter registration. I really appreciate them for the success that we had. I was happy. During the first shutdown, I was literally in tears. Just so many things seemed to be going wrong because of the pandemic. I remember saying, "No, we can't allow this to happen. There's no way that this is possible, we have to continue the work." And by golly, we did.

By November 3, we had four thousand voters registered and casting their ballot.

There was still a lot of setback where drop boxes were concerned. People were confused about which precinct they needed to visit to cast their ballots on Election Day. Early vote locations were open to anyone, but that still created a lot of confusion. The voter registration deadline was extended, and then it was shortened. Then it was completely stopped, so that created more confusion. We had the census going on as well. People thought that we were doing census work, so a lot of people turned us away. But, you know, it was exciting. It was trying. I was in an emotional state working anywhere from ten to twelve to sixteen hour days, driving six to eight hours across Navajo. I would set up in a location and be there for several hours to register voters. It was trying and exhausting, but it was well worth it.

I keep coming back to my people, to the community members, because they're the ones who are being left behind. They're the ones whose voices we need to elevate. They're the ones who need to heal. Forget the politicians, forget the leadership, forget all that and just be human. I remind myself to be human and to be humble. At times that has been a struggle, and I'm tested almost every day. We, as family, through K'é, could go so much further if we just put differences aside, be open-minded, and realize the injustices that have come and gone. We have the capacity to say enough is enough and decide what we want the future to look like for our children.

A better future has to start with education—whether in public school, college, universities, or just taking the opportunity to read a history book and ask questions. A lot of people will not answer your question if they don't have an answer to it, so you have to constantly question. If somebody gives you an answer, and it still does not sit right with you, take the time to ask somebody else who is knowledgeable in that area. Take the time to educate yourself, research. Take the time to educate your kids. Educate one another, give the person next to you that tool to carry forward and give strength to the rest of the communities and the next generation.

When we decide to start having conversations with the youth and the elderly, we need to give them the tools right now. Within my generation, my peers, I feel like a lot of us are lost because we've been given this opportunity to live on the reservation and also live in the city. We know both worlds, and I've seen a lot of my peers struggle when they first entered the urban areas. You could see and hear that fear. But it's okay to be scared. It's okay to take that fear and use it as a tool as well.

There's so many things that we could tackle by talking to one another and saying, "Regardless of what happens, I'm here for you. Regardless of our differences, I'm here for you." As long as we're willing to help one another, to educate one another, and to support one another's work, we can go so much further. Yes, it is very trying, especially in this time right now, where leadership has failed us time and time again, where treaties have been broken, where promises have been broken. But we still have the opportunity to continue to fight. We still have the opportunity to give a better life to the next generation. For that child that's going to be born in the next thirty seconds, for that child that's going to be born in the next year, we still have the opportunity to make change possible. We just have to find it in ourselves.

Yes, we have traumas. We have all experienced traumas in our lives. The traumas that I have seen growing up, at one point in time, I let it get the best of me. I let it get the best of me, and I faltered. I had almost given up, but your grandma and your parents tell you your prayers come hand in hand with what you do on a day-to-day basis. If nothing else, pray. Pray about where you want to go, where you want to be, who you want to be, who you don't want to be. Through my prayers and my belief in my Creator, look at where I am. It's possible for everyone else. We all still live through traumas. Look at what's happening at the Capitol currently. It's another trauma that we're having to experience. But we have overcome these traumas before. Our ancestors have overcome so much in their lives. We can stand together and truly believe in one another and be willing to learn.

Understanding the human nature of us all and listening to each other can help create a healthier future for everybody. What does it mean to heal? What areas are we trying to heal? Where? What are we wanting to heal? We know we need to heal Mother Earth because she's slowly dying and deteriorating. But what about healing within ourselves? Healing happens on so many different levels: healing ourselves, our spirituality, our physical being, our mental being, our emotional being, and then healing toward each other, forgiving one another, understanding one another. Right now, as human beings, we need to realize that we are only human but we still do matter. We're killing each other off. We're killing Mother, our Mother Nature, Mother Earth. How do we fix that? How do we go about fixing that when we can't see within ourselves what we need to heal?

I grew up with both traditional teaching and church teaching. My mom and my grandmother always said, "Prayers are the answer to everything. First thing in the morning, pray. To end your day, pray. But don't sit around and wait for your prayers to be answered. Make your prayers a possibility and make your prayers a goal for you. Set those goals within your prayer within yourself."

In order for us to heal, we need to find healing in ourselves and forgive ourselves. Despite where we've been, we need to forgive one another for the wrongs that we have done. The best thing to do is to forgive one another. I pray every day to forgive whoever has done us wrong, like the Trump administration. It is not for me to judge them, because in the end, they will be judged. That's where the traditional and the church teaching helped me. I was so confused throughout my youth about how to make both those teachings work. But now that I've done this work with my families and my community members, I have brought the two strings of each teaching together. To ask for forgiveness has brought me a long way to put my demons to rest, to face my traumas that I experienced on so many different levels.

Throughout history and the discovery of the Indigenous nations across this continent, Indigenous people have always been a problem. We needed to be removed, we needed to be assimilated, we needed to be converted. We were always that problem that needed to be resolved. When a person hears that from a delicate age, it causes them harm. We need to encourage and teach our youth who they are and where their stories started to ensure they have a stance when challenged or questioned about who they are.

Today's youth, they have so much knowledge within their fingertips with new technology. I tell them, take the opportunity to know your truth. Get it from an elder, get the true history of who you are and where you come from. In that way, we can start to bridge the gap with the elders and the youth. When we start to have conversations with one another, we start to heal.

Tara Benally is of Hopi descent, born for Navajo, and grew up in San Juan County, Utah. Her family was raised by the Diné Bítahníí clan (the folded arms people), but she was born for the Nítłachíí (marks on the cheek people). After being in construction for several years, in 2016 she became an advocate for her people with a focus on health care, education, community resilience, and economic development. Since 2018, she has served as the field director for the Rural Utah Project, working to register and turn out voters in the Navajo Nation.

FROM CHILDREN OF THE NÁÁTS'ÍILID, FOR THE DZIŁ ASDZÁÁN COMMAND CENTER

BY KINSALE HUESTON

Close your eyes, and let the land hold you.
Know the ways your body fits this earth—each slant of an eyelash,
Drape of a finger—
As it has always been,
This land holds its people,
And the people hold each other.

Sister, I see you,
With pain knotted deep like alien roots,
A garden of hurt that will not release you.
I see you sway against the rising moon,
In the juniper branch, in the rain sheet that promises
A resolution. Sister,

I hear you. The infinite breakings within your voice, the old ones
Brought to form each time you whisper to the land.
We have been taught to speak with a thousand voices,
And each day you remind me
How to find my words again. With them,
I build this path back home to you in offering. Sister,

I love you: the revolution that is your body, the soft earth
From which you formed. To heal, sometimes your hurt must split
Like an old wound to surface the stronger flesh.
We are taught to be caretakers by an ocean of beings

Who offer their blessings. You have only to reach out one hand to the
wind
To catch them on the rising dawn. Sister,

I know you as myself
And all this world has come to be; every cicada sings for you,
Every sky breaks open with rain for you, for you
Every peach tree in the garden preps the soil,
Every cactus fruit blushes and bursts, every milkweed
Laughs in the wind. Every landscape wiped clean with monsoon
Unearths the bones of small treasures for you. Do you not see
How the bluebird waits—
Perched on the edge of the heavens—
To bring you seeds and soft when the pain edges on unbearable?

And I?
I write this for you with every voice you have given me.
Not much, but how do you redeem all that sustains you?
I watch and love you, sister.
I will protect you.
And we will hold each other in the breaking

Of each new day.

Kinsale Hueston (Diné) is a junior at Yale University where she studies Indigenous poetics and literature as a healing, communal practice. She previously served as a 2017-2018 National Student Poet, and is the recipient of the J. Edgar Meeker Prize and the Yale Indigenous Performing Arts Program Prize for poetry. Currently, she is a First Peoples Fund Fellow and the editor of Changing Wxman Collective, an arts collective and literary magazine.

SURVIVING MONSTERS

BY PSARAH JOHNSON

Halfway through my forty-fourth orbit around the sun, I discovered an alarming truth. I was dismayed to learn that there is no magical point on the timeline when one stops being afraid of silly things. I have to wear socks to the movie theater if I'm seeing a horror film. Otherwise, something will grab me from under the seats. It doesn't matter how hot it is, I have to sleep with at least a sheet over me knowing this gossamer layer will protect me from the ax blows of a homicidal maniac. I still pole vault into bed to avoid a scalpel-wielding, "undead" toddler severing my Achilles tendon. I avoid the deep end of the pool for fear that a manatee will pull me under, a phobia which remains even after my college biology class taught me that a manatee's natural habitat is not a chlorinated pool. As for basements, all I have to say is I live in a single floor rambler, and I refuse to go to bed without something heavy wedged against the access panel to the crawl space, which I have never entered.

In the early months of 2020, I discovered that while one's childhood fears didn't necessarily disappear, they could be muted when confronted with a real monster. As a lifelong immune-compromised and chronically ill human being, I had always known I was at "higher risk." I didn't necessarily know exactly what that meant. I was informed one day by an ignorant classmate that I could die of arthritis. Other students reminded me of the "fact" that arthritis was contagious by chanting "leper" and running away if I got too close. Regardless, my mother and rheumatologist reminded me that these children were being a tad hyperbolic. So, dying didn't seem like a very likely outcome for me. Certainly not as real a threat as the troll that lived in the shower drain and would cut off my toes if I stepped on it. Death, even factoring in my compromised immune system, could be easily avoided. At least such was the case before March of 2020 and the onset of COVID-19. My imaginative childhood fears slowly began to recede as new and very real fears emerged.

It began slowly. Initially, I learned that we could avoid the virus by quarantining. This was certainly no biggie to me. I, along with a large swath of the disabled population, knew how to do that! I had done that every flu season, especially when taking immune-suppressing medications. Disabled advocates, being well-versed in how to quarantine, how to keep a sanitized living space, how to mask and glove up when leaving that bubble, were the first to step up with protocols on how to stay safe. Zoom meetings and telecommuting became standard operating procedure, though we'd been asking for these "reasonable accommodations" since Bill Gates made his first million. We began mutual aid networks across the country, started collecting, inventorying, and distributing PPE. All while the Narcissist in Chief downplayed the crisis, devaluing us and our frontline efforts as expendable, casualties in the fight to save the economy. And his rabid fans soon took up the cause, openly admitting that these casualties were an acceptable price to pay to protect our American culture of corporate greed and wealth worship.

Then some of my progressive friends joined in, deciding the best way to ensure their own financial security no longer lay in social programs funded by a tax on legacy billionaires and financial corruption. Suddenly they too were happy to accept the death of a few for the good of the many; not the good of many lives, mind you, just many paychecks. The fear of losing their economic security was every bit as real a fear as COVID-19 but their response was akin to sending Godzilla to defeat Mothra. The destruction created by this so-called solution, re-opening the economy too far and too early, did nothing to solve the underlying problem. People began attending the anti-lockdown rallies, encouraging reopening businesses while the virus still ran amok. Many claimed masks were causing health problems, some even trying to cite the ADA as they claimed going maskless was a "reasonable accommodation." This campaign, bred in ignorance, led to an infection rate higher than ever and a growing acceptance of anti-mask sentiment.

Soon I found myself unable to sleep. I sat on the edge of my bed, both feet planted firmly on the floor. I would roam my dark house, "Naked and Unafraid." When daylight began to seep through the slats of my blinds, THAT was when I ran for the sanctuary of my bed and its cottony layers of protection.

The simple act of leaving my house felt dangerous. I felt like the lamb, with my unmasked neighbors—wolves licking their chops. The wheelchair constitutionals I used to take around my neighborhood took on a Mad Max feel as I would have to choose between veering into traffic or wheeling past an unmasked asshat defiantly coughing as he passed. I felt my odds were better in traffic—until I began to hear stories of the Trump trains intimidating folks for wearing masks. Then there were the anti-mask shoppers flouting the safety protocols established by grocery stores with the input of local governments and medical professionals. On one trip to the grocery store (during elderly and disabled early shopping hours), I realized I hadn't been keeping an eye on the time. Just as I was at the end of the cereal aisle a mask-less man turned the corner and started toward me. I asked if he'd wait until I cleared the aisle, just another four feet. He brazenly kept coming. I backed my scooter down the entire length of the store, out to my car, and never went back. I began to feel like a target in more ways than one: A) I was one of those goddamn maskers that hated "Murica," B) I was a fucking cripple, and C) my bald head revealed me as a cancer survivor who had the audacity to think I deserved to live through this pandemic.

Yes, in true 2020 fashion I had discovered a lump in my breast the size of a walnut. On April Fool's Day! By the twentieth, I was undergoing chemotherapy and by May Day I was as bald as a cue ball. At this point, my childhood monsters were completely defeated by very large, very real, and seemingly undefeatable fiends that stepped boldly out into the sunlight and left me no safe harbor. My neighbor, who only weeks ago had offered to help my husband and me by mowing the grass and bringing meals while I fought for my life, now spouted eugenicist theories of "herd immunity" and "survival of the fittest," while Darwin's bones rattled in his grave. I could hear him shouting through the veil, "That's not what I meant!" Ayn Rand has co-opted Darwin's theory of evolution to uphold a disgraceful and selfish notion. While Darwin's theory referred to the ongoing success of a species, Rand framed it in terms of the individual and who deserves life.

As ICUs began to fill, there was open debate over who deserved care and whether or not the chronically ill should be sent to the back of the line. I watched this debate play out from my hospital bed at the Huntsman Cancer Institute. I had not responded well to chemo, or rather had responded too well. My neutrophils, a type of white blood cell, were at zero. I literally had no immune system. I spent the loneliest week of my life (there was a zero-visitor policy in effect) listening to pundits infer that I was selfishly taking up a bed that could go to a normally healthy and productive member of society, temporarily "inconvenienced" by COVID-19. It was a rough week. Even more so for my husband and mother. Unable to stay with me or even come up for a visit, my family and friends were justifiably worried. In my neutropenic state, a fly with a cold could have finished me off. Luckily, my neutrophils returned. A week of careful monitoring and postponed chemo, and I was able to return home. But I could understand how it must have felt for those hospitalized with COVID-19.

My US House Representative, Ben McAdams, had been hospitalized with COVID-19 and he had shared what it was like. The pain, exhaustion, and weakness he experienced made neutropenia sound like a grand day at the beach. Yet regardless of our social stations, diagnoses, and prognoses, we were equally deserving of the care we received. But what would have happened had we simultaneously arrived at the doors of a hospital that was over capacity? In a decent society run by compassionate and intelligent leaders (which we had lacked for four years), this question wouldn't have kept disabled advocates up at night fearing the true monster in their closets: ableism.

While we worried over whether or not we would receive treatment at hospitals, an anti-science administration went after our ability to access our treatment at home. I had relied on hydroxychloroquine for ten years to keep my joints lubricated and loose, and to ease my pain. Suddenly, after an ignorant and unfounded suggestion that hydroxychloroquine could be used in treating and even preventing COVID-19, I found it nearly impossible to fill my prescription. For forty plus years, I had used the same mom and pop local pharmacy. I was on a first name basis with the staff, and yet their hands were tied. They simply didn't have the supply. It took calls to Walgreens, CVS, Rite Aid, and Costco before we could find a refill. This obstacle course was run monthly from March to July before supply trains recovered. Forget about toilet paper shortages, the hydroxychloroquine shortage of 2020 kept thousands immobilized, like trying to run a car without oil.

Still, I counted myself one of the lucky ones. With my parents retired and my husband largely working from home (thanks to COVID-19), I was able to stay in my house for the majority of my cancer treatment. For entirely too many disabled folks who lived alone, any additional medical emergency meant treatment at home was not an option. Instead, nationwide, People with Disabilities (PWDs) were being checked into nursing homes, assisted living facilities, and rehabilitation programs, and entirely too many of them were being left there, at the epicenter of the COVID-19 transmission. As if that wasn't enough, many lost their apartments during that period of institutionalization, as frequently their rent was rerouted to the institution leaving them essentially homeless. This is because here in the US nursing homes enjoy an advantage called the institutional bias where states receiving federal Medicaid dollars MUST provide nursing home services while Home and Community-Based Services (HCBS) are optional and routinely denied.

ADAPT National, one of the top disability-led advocacy groups in the country, had been fighting against the institutional bias for decades by asking that Long-Term Services and Supports (LTSS) be made available through HCBS. Under our current system, a disabled individual recovering from surgery or illness could recover at home only if their state Medicare/Medicaid program allowed for HCBS.

The Supreme Court decision Olmstead v. L.C., one of Ruth Bader Ginsburg's defining cases in 1999, found that "unjustified segregation of individuals with disabilities was a form of discrimination under the Americans with Disabilities Act (ADA)." While this decision made an important statement about the rights of PWDs, ensuring this decision was honored by all states has remained problematic, as is evidenced by the institutional bias still rampant in many states.

A hopeful solution exists, though. The Disability Integration Act (DIA), which has been introduced in the Senate, would change that, by giving, for the first time, PWDs their full rights to life, liberty, and the pursuit of happiness. The DIA would enable any disabled individual in any state across the nation to choose whether to remain home with in-home care or consider institutionalization. Of course, this Act makes the nursing home lobby nervous. Suddenly, money that had been earmarked for institutions, which ensured them a steady inflow of both capital and "customers," would now also be used for in-home care.

Just a week before Utah's March lockdown, ADAPT Utah, the advocacy arm of Disabled Rights Action Committee (DRAC), sat down with Senator Mitt Romney's staff to discuss the need for the Disability

Integration Act. In the time of COVID-19, the number of cases and deaths in these institutions were distressing. Nursing homes were considered hot beds for the virus. The threat to the staff and patients of these institutions became a national concern, yet many of these patients, overwhelmingly PWDs, didn't need to be there but had no financial alternative. While Senator Romney never got back to us with a definitive answer, Senator Lee's office and our House Representatives hid behind the "state's rights" excuse claiming each state could decide for themselves, leaving disabled folks even more exposed to COVID-19 because nursing homes had monopolized Medicaid dollars. In fact, not a single Utah Congressperson has so far co-sponsored the DIA. Regardless, the DIA was introduced January 15th, 2019, with bi-partisan sponsorship at a time when the divide had never been wider. Together, Senator Chuck Schumer (D-NY) and Senator Cory Gardner (R-CO) introduced the bill to the Senate, and Representative Jim Sensenbrenner (R-WI) introduced it in the House. The DIA had thirty-eight Senate co-sponsors and 238 House co-sponsors yet it sat ignored all year by the 2020 Congress at a time when in-home care was more needed than ever for People with Disabilities. As the US was told to shelter in place, this community was left with no good options. The message heard by disabled activists was that our existence continued to mean little.

For far too long the term disability has been stigmatized as weakness, a painful reminder of one's own mortality. I believe that this characterization is not only false but also harmful to humanity as a whole. A society that devalues disability and fears pain is a society that does not value survival. A society that chooses to hide people with disabilities from the mainstream is ensuring that people are wholly unprepared if and when they become disabled or chronically ill (as has happened to many post-COVID-19 patients). When our media fail to show disabled folks telling disabled stories, that media is responsible for keeping our society ignorant, unprepared, and negatively prejudiced. When our media take the time to show disabled bodies and listen to disabled voices, then society as a whole benefits, and stigma and fear are replaced by acceptance, understanding, and empowerment.

It is my sincerest hope that as we give the boot (and the finger) to 2020 and America's most incompetent administration, we will move forward having learned some valuable lessons in what NOT to do in the midst of a pandemic:

- Don't give a man-child with no medical background the microphone.
- Don't ignore the advice of medical professionals.
- Don't treat science as a belief system that you can choose to follow (or not).
- Don't fuel your economy by throwing the disabled, chronically ill, and elderly into the furnace.
- DO involve people with disabilities in the conversation on surviving a pandemic. There is no other group as well versed in the simple subject of survival as People with Disabilities..

Maybe if we follow these simple suggestions, we can slay the COVID-19 dragon and I can go back to my mundane fears that a spider will lay its babies in my ear canal.

A self-identifying "cripplepunk," Psarah Johnson was born with juvenile rheumatoid arthritis and has spent her life collecting autoimmune disorders. She also suffers from PTSD, a result of medical trauma and psychological abuse. Psarah has worked as a group home coordinator, taught high school students theatre and civics, performed as a stand-up comedian, and—at one point—even ran away to join the circus. Psarah is now a full-time activist, advocate, writer, and public speaker and serves as board chair for the Disabled Rights Action Committee.

FILLING A CRUCIAL NEED
VOLUNTEERS SERVE ON FRONTLINES DURING PANDEMIC

BY ALASTAIR LEE BITSÓÍ

The Team Rubicon Greyshirts, like Amy Shields, love serving the Diné people with nursing care and have been volunteering on the frontlines since the COVID-19 pandemic began its surge across the Navajo Nation. Team Rubicon, a nonprofit made-up of 120,000 Greyshirt volunteers—military veterans, first responders, and civilians—deploys help to communities hit by natural disasters and humanitarian crises.

"I gotta go, I gotta go. Period," Shields, an emergency care nurse, remembers saying when the Navajo Nation reported its first COVID-19 case in the Kayenta Service Unit in the spring of 2020.

COVID-19 has since killed over one thousand people of all ages across the Navajo Nation. In the Kayenta Health Center emergency room, Shields treated thousands of COVID-19 positive patients and saw as many as eight patients per shift transported to larger hospitals outside Dinétah. In the first surge, Shields did her best to comfort Diné.

Some of those crushing moments, she recalls, saw her standing in PPE and scrubs watching loved ones say goodbye behind glass doors. Some family units had ten to fifteen members infected with COVID-19. One by one, Diné people came for COVID-19 related issues with many thinking it was a death sentence because of the number of fatalities from the virus, Shields said.

"We were there during the tough times," she said. "We had patients come non-stop."

In the first surge, Team Rubicon sent 128 Greyshirts, including 47 medically trained volunteers like Shields, to the Navajo Nation between April and June of 2020. In total, Greyshirts serving the Navajo Nation clocked in over 15,000 hours in more than 90 days and served approximately 3,206 patients. As of spring 2021, the Navajo Nation had over thirty thousand cases of COVID-19, with over 1,200 deaths.

Because COVID-19 is highly transmissible and has shown high fatality rates among the Navajo Nation's 160,000 citizens, the Navajo Area Indian Health Service, Nez-Lizer administration, and Navajo Department of Health asked Team Rubicon for relief to help mitigate the virus's spread.

Dr. Lorretta Christensen, chief medical officer for Navajo Area Indian Health Services, asked volunteer organizations like Team Rubicon for help. Team Rubicon was the first volunteer organization deployed to the Navajo Nation, specifically to the outbreak zone in the Kayenta Service Unit. "They're definitely filling a crucial need," Christensen said. "We were so busy at every health facility with the second wave. We are so happy they came back again. They're very knowledgeable at public health emergencies, like logistics and testing operations."

Christensen added that volunteer organizations, including Team Rubicon, are critical to alleviating the shortages of medical personnel across the Navajo Nation. Other volunteer organizations assisting with epidemiology, contract tracing and medical care include Community Outreach Patient Empowerment, Johns Hopkins Center for American Indian Health, Doctors Without Borders, Global REACH, Veterans Administration nurses, and the University of California-San Francisco's HEAL Initiative, among others.

After COVID-19 cases begin a downward slope during summer 2020 and then rose in the fall, Team Rubicon deployed its second mission in November 2020 to the Gallup Indian Medical Center after being asked by Navajo Area IHS to help during an increase of COVID-19 related hospitalizations. Greyshirts assisted in emergency rooms and at testing sites.

Team Rubicon Greyshirt Terri Whitson, a nurse and Navy veteran, administered the first rounds of COVID-19 vaccinations to GIMC medical staff on December 15, 2020. Team Rubicon has since helped the Navajo Nation during its mass vaccination effort.

This is the first time that Team Rubicon has served Indigenous communities, says Dr. David Callaway, Team Rubicon's chief medical officer. While he has provided care on the frontlines in North Carolina, Callaway said Team Rubicon is happy to help the Navajo Nation with COVID-19 medical relief.

"The way that we deploy is at the request of the Indian Health Service," he said. "Even as our mission ended, we stayed in close contact with them. We work with local health systems, and come up with a plan to most effectively collaborate with the team on the ground."

For Shields, who has been on the front lines with other Diné doctors and nurses, it is her connection to Dinétah and growth as a nurse that makes her mission under Team Rubicon a natural fit.

Or, as she put, her belly button is buried in the Navajo Nation because it is where her path as a nurse began. "The Navajo Nation is near and dear to us," Shields said, adding that her daughter attended school in Fort Defiance. "When all this happened, I knew I had to get back there. I literally grew up on the Navajo Nation as far as my career is concerned. I know how to say, 'Ya'ateeh," she said. "Shi ei Alicia yinishye."

Biography on page 263.

YOU'RE DINÉ BEFORE YOU'RE A CONSERVATIVE

BY ARLYSSA BECENTI

In our Diné Creation Story, Diné (Navajo) people were created from Changing Woman. Our four original clans come from her as well. Throughout history, our Diné women have been the strength of Dinétah. During the Long Walk, when Diné prisoners were force-marched from our homeland to Ft. Sumner, New Mexico (Hweeldi), it was Diné women who had played an integral role in our people's freedom to return home. These are only a few examples of the importance of Diné women's role within our society, or, as I like to say, Indigenous femininity rooted in matrilineal teachings.

As a Diné woman, who was raised by a strong, fierce, outspoken, intelligent mother, and a soft-spoken, kind, loving father, I was taught to be genuine in nature, empathetic and compassionate toward others, but to also not be a pushover. I've learned my father is unique with matrilineal teachings: he is close to his older sisters and sisters-in-law, has enduring love for his late mother-in-law, and is devoted to my mom and two older sisters. I thought all Diné men were like him.

In the virtual world of Twitter, this is not the case. I check my Twitter page, and, no surprise, my recent anti-Trump tweet gets a distasteful comment from another conservative, who sadly happens to be a Diné man. No problem. I'll read the comment and decide whether it's worth responding to, and I realize this one is not.

I'm a Diné woman, and a journalist for the *Navajo Times*, my coverage has mainly been on the Navajo government. Throughout my career I've learned how toxic masculinity continues to be a theme that I experience when trolls come at me about my articles, and how they are written. When they read my personal and political views on Twitter, especially when they see I am a critic of toxic masculinity and white supremacy that Donald Trump represents, their dislike of me becomes increasingly loud.

Throughout 2020, I spoke against this on Twitter. As a reporter I have to be unbiased and balanced, but on my beat I do not report on Trump. When I do mention Trump in my articles, it's nothing like my rage-filled personal tweets against him. Even so, Trump followers, Navajo or non-Navajos, don't read the *Navajo Times*; they read my tweets.

What Trump and the Republican Party have done is shine a light on Diné men who follow the ideology of "Guns. God. Trump," which was on a shirt I saw a Diné man wear during a Trump rally in Gallup, New Mexico.

I use my Twitter account to update recent COVID-19 case and death numbers for the Navajo Nation, and other related coronavirus news. This daily update is something I've been doing since March 2020 when our Nation received the first report of COVID-19 in the Navajo Nation.

Initially, my whole point of Twitter was to be a source of information for urban Diné or Diné who are no longer living on the Navajo Nation but want to keep updated on Diné news. As the government reporter for one of the most influential tribal governments, I felt it my duty to keep readers up to date on what their leadership is doing and discussing. Twitter in 2020 was also a source for people around the world who were interested or concerned about the high prevalence of COVID-19 infections in the Navajo Nation.

"It would be great to get Trump supporters just as passionate and concerned about COVID as they are about their votes. Extra points to have them believe science as much as they do their lying President and crazy conspiracy theories," I tweeted after the election in November 2020. This is around the time the Navajo Nation was heading into its second wave of COVID-19, and the rest of the country was going into its third.

I knew Trump was popular, but I didn't expect that Diné men would go out of their way to try to insult me and put me in check. I do not feel forlorn, but rather I feel sorry for these men who feel it necessary to pick fights with a strong and resilient woman they don't know. Do they speak to their moms, sisters, aunts, grandmothers, nieces, and cousins like this? I could be related to these men by clan. Yet, they attacked me for attacking their president, as though I was attacking them.

Somewhere in their life these Diné men lost part of their Diné way of life and thinking.

This is only one example of misogyny I have encountered on Twitter. At the beginning of summer 2020, I was covering a huge story of illegal hemp/marijuana growth in Shiprock, Navajo Nation, New Mexico, which was spearheaded by a career politician named Dinéh Benally. In this beat, I found that I was up against Diné men who agree with how Benally broke Diné law for his own gain, exploiting our own lands and own people while in the process. I received other harassing Tweets about this and was made into a meme, along with my coworker and photographer, Sharon Chischilly, because of our continuous coverage.

To add insult to injury, a white male reporter had covered the same issue, not as extensively, and he was praised for his reporting and never made into a meme. In his coverage he never gave me or *Navajo Times* credit. Instead, his white male coworker claimed they broke the story, even though I had written half a dozen articles before his first article came out. I was furious!

Being underappreciated, scooped, and ridiculed by men all because

I felt it my duty to keep my Diné people informed about COVID-19, the illegal hemp farms, and Trump's fascism, racism, and hatred, was frustrating. Were my ideals for living up to what a strong Diné Asdzaa (Diné woman) is supposed to be for her people unrealistic and foolish?

As a Diné Asdzaa, we are the strength of our people, so my ideals and what I am trying to accomplish isn't foolish or unrealistic. It's what is expected of me.

Unlike my dad, who is a follower of his Diné cultural practices, conservative Diné men seem to have shunned our tribe's matriarchal society. Trying to fit in with the colonizers and their patriarchal, Christian, capitalist ideals, conservative Diné men are quick to scream "fake news" at the media that doesn't align with their Trump love.

They have no qualms at harassing me as a journalist and a woman. But they forget I am not just a woman, I am a Diné Asdzaan, and like all other Diné women I have power, strength, determination, intelligence, whether our own men like to admit it or not.

These conservative Diné men I've angered and annoyed are set in their ways. Their paths will continue to align with a racist, misogynistic narcissist. But, there is one boy I can teach to treat women respectfully, to follow the path of Diné teachings and culture, and that boy is my two-year-old nephew.

I want my nephew to grow up in a world that isn't ruled by a capitalist, racist colonizer, where his homelands aren't damaged by oil and gas extraction, where he can breathe in clean air, where sovereignty rights means something, where he can learn his culture, language, and traditions and keep them alive. Above all, I want my nephew to grow up and to be a good relative to all the natural beings in the world, and to respect women, just like his dad, and his Ánálí Hastiin (paternal grandfather, who is my dad).

For some Diné men, being good and respectful toward Diné women won't happen and whether that will change is up to them. Maybe they have been hurt by women before and see me as a target to take it out on. I don't know their stories, just like they don't know mine. But I say this to them: if you're fueled with this much anger at me then get angry about things that matter. I am one woman, but there are multiple unjust and terrible things happening in the Navajo Nation that need your anger to change it for the good. Return to Diné teaching and beliefs, where people, land, and animals are all a priority. Greed, hate, selfishness is what Trump and conservatism is all about, and that's not Diné.

Arlyssa Becenti is a Diné journalist originally from Fort Defiance, Arizona, and she's an alumna of Arizona State University. She has reported for Navajo Times *since 2016; before that she reported for the daily* Gallup Independent. *At both publications she mainly reported on Navajo Tribal Government. In her decade-long career she has won awards from the Arizona Newspaper Association, New Mexico Press Association, Native American Journalism Association, and Arizona Press Club. Her clans are Nát'oh Diné'é Táchii'nii, Bit'ahnii, Kin łichii'nii, Kiyaa'áanii.*

WHAT WE KEPT

BY LINDA HOGAN

We had mountains
and you took down the trees
so that rain felled the mountains.

It was once enchanted
with the song of golden winds,
the silk thread of river,
pollen from the medicine flowers
you took. The whole world
was the gold you wanted.

In the past, we gave you our labor.
We gave you our store of food,
even the mats where you slept for a year
before we sent you away with burning arrows
and your fat ran across earth.

You took the plants
from our beautiful woods
on your ships
to lands already destroyed,
and even more of you arrived
to take our homes
while we still lived inside them.

You took the birds
from the rookeries of beautiful waters,
feathers for hats from near the mangroves,
coats made from our animals,
and all the time

you lost so much, even taking,
because you knew so little
that a girl led the way
to the fame of men,
fed them, turned their canoes
to safety.

The more you took,
the more you lost.
And you need us now
the way
you needed us then, our land
and labor,
and we give to you
knowledge you don't hear,
the new mind you can't accept,
our bone and leaf soup.

But what I keep
to myself,
for myself, is the soul
you can never have
that belongs to this land,
the magic haunting you
still and always untaken,
but you want,
how you want,
how you need.

Biography on page 12.

ANEEST'Į́Į́'
AN INDIGENOUS ARTIST'S PERSPECTIVE ON THEFT AND RECLAIMING

BY DENAE SHANIDIIN

As an Indigenous and Korean artist in a colonized society, I exist along with my relatives in environments where the theft of our cultures has been integrated in every part of society. White artists, corporations, and other institutions of white supremacy have coopted social justice movements. This is done using the tragic deaths and brutalization of BIPOC victims of white supremacy and colonization under the guise of honoring or memorializing. Most of us have become familiarized with the beautiful faces of Breonna Taylor and George Floyd, and I have questioned how we are guided as artists. How and for whom is human rights and environmental art made? Who is granted access to create that work? Who gets paid for doing this kind of work? Who has a right to tell these stories? Who funds it? How will the work communicate an array of people of different nationalities and identities?

I have been walking and growing alongside Missing and Murdered Indigenous People (MMIP) work as it has been expanding in media attention since 2019 with the Presidential Task Force on Missing and Murdered American Indians and Alaskan Natives (Operation Lady Justice), Savanna's Act, and nineteen other bills within the so-called states addressing the crisis. But I can say with confidence, at this time, there is little to nothing being done to actually protect Indigenous People from being victims of homicide, sexual assault, homelessness, domestic violence, and trafficking. In fact, it worsened with the COVID-19 pandemic.

Over the last few years, I have been learning how to navigate, educate, and create campaign awareness for MMIP through MMIWhoismissing, an Indigenous-led platform for MMIP action and voice. We advocate for grassroots efforts and support tribal coalitions laying the groundwork, socially and politically, to protect our Indigenous people from further colonial violence that leads to MMIP. I was guided into this work through collective, historical, and intergenerational trauma. As an amá yázhí (a little mother), a daughter, a sister, and as an adzáán (a Diné woman), I communicate and understand this violence in the context of the murder of my late Auntie Priscilla when she was nineteen years old. To do this work in a good way, I know I had to be on a pathway to healing myself as a relative.

I question in what ways I and others are responsible as artists for voices and stories of such trauma, resilience, and pain our relatives experience within colonial occupation. For me, making MMIP clothing and merchandise never felt right to do. One, because I find it personally harmful to my body to wear clothing items that carry a deep amount of pain, and two, to mass produce awareness and campaign merchandise of Missing and Murdered Indigenous People could exploit BIPOC suffering by participating in facets of capitalist gain. However, Indigenous people are incredibly resourceful and often must participate in the very system that dehumanizes them, as a strategy for survival. I am not critiquing Indigenous folks who do create MMIP clothing or other wearable items to bring awareness, as it can be empowering to embody the recognition of a very real and serious issue in our communities.

Being forced to participate in capitalism does not mean you are a capitalist, it only means that we live in a capitalist society, which makes it critical for us to understand that an anti-capitalist approach to restoring sovereignty and freedom for BIPOC allows the BIPOC community full autonomy and control over their own narratives. If honored in a good way, the education provided by these sacred stories should protect these narratives from white gaze, appropriation, and profit.

Settler society's false stories of Indigenous people fuel extreme violence. The misrepresentation, dehumanization, and sexualization of Native Women has contributed to the ongoing missing and murdered Indigenous people crisis. The stories of young girls such as Sacajawea and Pocahontas (not even their real names) were made popular by Hollywood and colonial narratives for the white gaze. Their true narratives carry symbolic and real trauma of trafficked Native women and girls. Indigenous people are at a loss of resources to combat this rampant violence, violence that rose during a hateful administration, in a time of climate catastrophe, and during a global pandemic that has measurable disparities in resources and disproportionately high death rates among Native populations.

The way that I personally navigate creation of art and content within social and political activism is to operate in a way that abolishes ownership, even as an artist. This is the opposite direction of what I was taught in art school. I believe the creation of work and messages that I produce are not ideas that I own, nor do I desire to own them. I know that ownership is rooted in colonialism, property, and whiteness and to make this work spiritually sustainable, I cannot operate from a place of individualism. The political messages and affirmations of truth and humanity that I produce do not belong to me, nor should they belong to any single person. My role as an artist is to learn how I can best be a meaningful relative. I must be a learning relative who understands that the work I produce affects my nation, my community, my clan, and my family. Ideas that allow us to operate from a place of sacredness and beauty are most difficult to bring into a settler society.

As so many social justice issues have converged, I've found myself sensitive to optical activism. I've seen Indigenous and non-Indigenous people take advantage of movements for their own monetary gain and personal clout in gallery spaces, in the production of murals, the selling of prints, and on social media. This weakens our perception of real change.

For the sake of understanding the predatory nature of white artists involved in BIPOC social justice issues, I will use an example from my community. In the summer of 2020, I decided to post my concerns about white muralists and the co-opting and appropriation of BIPOC stories. Josh Scheurman, a Salt Lake City, cis-gendered, white, male mural artist assisted in the painting of a George Floyd mural along with other BIPOC murdered by police. These murals have now become an important community gathering place. Initially the murals were undertaken by white artists as to not put BIPOC at harm and the muralists intended to remain anonymous. However, when Scheurman posted about the mural and took credit on his Instagram, this led to a false ownership over the creation of the mural. He was praised for his character, his "activism," and his artistic involvement.

Scheurman has a long and steady history of culturally appropriating Indigenous iconography and culture while centering white environmental voices, like his own, above those of Indigenous environmental activists. This is a common behavior that permeates mainstream environmental discussions regarding land and is a tactic rooted in Indigenous erasure. A couple years prior to the George Floyd mural, Scheurman painted a mural of Bears Ears, a sacred place for my Diné people and other Tribes, across the street from a popular brewery in Salt Lake City. When I saw the mural, I had fire in my tummy because the white environmental and art community had again co-opted Indigenous ancestral lands as their own. It is the same spiritual energy of the extractive industry, creating deep wounds within the Indigenous Relatives who have ancestral ties to that land while experiencing the most harm in a world of violent white supremacy.

Scheurman is not at all an anomaly. His behavior is celebrated widely and his art is encouraged and supported by those who share the same acceptance of toxic settler behavior instead of operating in a way that supports artists who are inherently tied to the subject matter. For example, if an Indigenous artist had been engaged to paint the Bears Ears mural, that Indigenous artist would infuse the work with an accumulation of thousands of years of knowledge sharing and five hundred years of resilience regarding the degradation of Indigenous land and body. Instead, communities continue to support the colonial epitome, cis-gendered white men to tell the stories of Indigenous land. Scheurman represents the opposite history from those who carry the true stories and spirit of Bears Ears, whose DNA is rooted within the land, way of life, and thousands of years of creation and beauty.

As I reflect and mourn BIPOC murdered during 2020, and see many platforms, individuals, institutions, and communities create honorary art of those wrongfully taken, I feel unsafe. During the many times I have pondered my own life being at risk of murder as an Indigenous woman, I have envisioned how the world would memorialize my death in a context where there is no justice for murdered Indigenous women. I am horrified by the possibility of a Josh Scheurman using such injustices to prove his "activism" for ego and social gain.

As Indigenous people, when we make art it is alive. Our histories, cosmologies, and identities are often preserved through art making. The modernization of art has not removed the traditional practice and spirit of storytelling and knowledge sharing. Much of why we create art as Indigenous people has shifted into healing our nations and acknowledging resilience, restoring traditional practices while also participating in spaces that have erased us, such as cities and galleries. By approaching artmaking as sacred, as prayer, we are reclaiming our identities in resistance to a long history of cultural genocide and forced assimilation. When society not only allows but supports the theft of our spirit in art practices, it creates deep societal confusion about who we are as Indigenous people. This theft simultaneously perpetuates the colonial narratives of who we are. It supports the predatory nature of non-Indigenous artists as people who continue to challenge our ownership of very specific symbolism, iconography, and ceremony that has evolved within our tribes for centuries.

These discussions are not just my own personal critique but are shared by those in my community who also experience the harms of white supremacy, which led to a group of BIPOC artists, activists, and individuals contributing to a set of understandings. Although these "understandings" were created out of a deep-seated problem in the mural community of Salt Lake City, Utah, they were learned through personally experiencing the spiritual and social implications of cultural appropriation, BIPOC suffering, and the colonization of land, body, and spirit.

The intention of these understandings is to create collaborative worlds that are just, equitable, trauma informed, victim-centered, anti-racist, anti-appropriative. In doing so, artists become more powerful and can have impact in dismantling colonization, gentrification, white supremacy, and racism in our communities.

We are all experiencing a collective trauma, some far more than others. It is a growing pain that propels us to move forward as active listeners. It is long past the time where artists must deeply consider how we are creating and who we are creating for in our communities. Art alone cannot fix the profoundly and deeply embedded racism that flourishes in every aspect of our lives, but artists have always played a pivotal role in the culture we intend to foster. Artists have a deep responsibility as relatives to address many of the harms past and present that have been adopted by many of us, particularly with art for social and environmental justice.

There is not much left that hasn't been stolen by white supremacy or convoluted through capitalism and hoarded by way of invasive powers.

However, despite spaces where spirit has been blanketed with whiteness, Indigenous peoples, Black and brown relatives, and People of Color remain the experts of their own realities and histories. We carry these stories within our bodies and they are passed down to our children with the understanding that they will be carried into the many generations to come. I truly believe the solutions lie within learning to be a good relative.

These "understandings" represent some community-based knowledge of respect and kinship. These understandings have no author and can be embodied by anyone who wishes to guide their own artmaking, activism within community, or changes of ethics within institutions.

We understand that Art as voice for public consumption needs to be treated with the utmost care and respect.

We understand that when murals and artworks pertain to social and racial justice issues, those directly affected must be centered and play a critical role in the process.

We understand the murals play a role in the gentrification of BIPOC and impoverished communities and conducting artwork must be for the people by the people, who represent and have historical identities tied to those specific communities and land.

We understand that the appropriation of different cultures and sacredness is a real issue, that it is rooted in colonialism and theft, and that it further perpetuates harm on the rightful owners of that sacredness.

We understand that the BIPOC community has the autonomy and control over their own narratives and that the education provided by these stories should protect these narratives from white gaze, appropriation, and profit.

We understand our commitment to putting BIPOC in power and centering them in the decision-making process.

We understand that art for social/environmental justice will not be performative for popularity or white clout; the art we make will have intentions to make real change.

We understand that paying or compensating BIPOC including queer, trans, and disabled artists for their time and counsel is being in good relation with those relatives.

We understand that honoring those killed, lynched, raped, and murdered will be memorialized with the utmost respect, sensitivity to family and community, and will center the appropriate parties connected to the individual's unjust absence.

We understand that environmental justice and Indigenous sovereignty are interconnected; one cannot exist without the other and those voices must be at the forefront.

We understand that the social profiteering from our movements is optical activism.

We understand that clothing items for social justice will not profit off of specific issues of BIPOC death and suffering.

We understand that the members of our community will often times need to be called in to address problematic and racist behavior; addressing these harms is an act of love.

We understand that naming our oppressors is an act of justice, and that justice requires community care and protection for the victims of the oppressor.

We understand that asylum seekers must have access to public art opportunities and beyond that provide them with community inclusion and support, honoring their artistic intelligence brought with them.

We understand that the social media of institutions and individuals often do not reflect the reality of their ethical practices.

We understand the importance of addressing barriers to accessibility when it comes to the physical mobility of an artist, through supporting any physical assistance an artist may need to fulfill their visions in public art.

We understand that fighting to be right as a white person when being addressed by BIPOC instead of seeking to understand is white supremacy and is an act of oppression.

Denae Shanidiin, a Diné and Korean artist, is born to the Diné (Navajo) Nation. She is Honágháahnii, One-Walks-Around Clan, born to the Korean race on her father's side. Kinłichíi'nii, the Red House People is her maternal grandfather's clan and the Bilagáana, White People, is her paternal grandfather's clan. Shanidiin's projects reveal the importance of Indigenous spirituality and sovereignty. Her work responds to her identity as an Indigenous woman and artist and brings awareness to many contemporary First Nation issues including missing and murdered Indigenous people.

AN ILLUSTRATOR'S GUIDE TO GRIEF

BY MARIELLA MENDOZA

I will adorn you with color and light
-color and light
-color and light
and if that's not enough I will add more color
shades of blue to accentuate your beautiful brown skin
different flowers to remind me of your essence
soft hues, depth and shadow, a dance between what was gone and
what
could've been
and if that's not enough I will add more light
stars around your body
carrying your memory via constellations
impossible to describe what you meant to me
every single day I dream of you
and if that's not enough I will demand of the world
to not forget your beautiful face
to not forget your beautiful life
I will add big bold letters, to remind everyone
what was taken from us
I will add all your favorite colors, to remind everyone
that you were our kin
but most importantly
I will adorn you with color and light
not because I refuse to move on
not because I refuse to sit still
but because you moved me

shaped me
changed me
and you, beloved, live in me.

Biography on page 56.

WE ARE STILL HERE
LAND DEFENSE IN THE BORDERLANDS

BY NELLIE JO DAVID

As told to Alastair Lee Bitsóí and Brooke Larsen
January 2021

I'm from Arizona. I'm from a mixed heritage. I'm Tohono O'odham, Hia-Ced O'odham, a mix of European heritages, and Mestizo, and I am from the borderlands. Ajo is about a forty-five-minute drive to the Lukeville Sonoyta border area. Historically, Ajo is Hia-Ced O'odham land, which is on the west side of O'odham territory. O'odham land extends as far north as the Salt River up in Scottsdale, as far east as the San Pedro River in Tucson, as far south as Caborca, and as far west as the Yuma territory. And maybe even beyond that. In the era where I've grown up, our traditional territory has been divided by the border and has been impacted by the ever-increasing amount of militarization. I never intended my life to be on this platform speaking about militarization or fighting for the rights of our people who are occupied by the Border Patrol and by various forces surrounding us.

My dad's an artist. My family's very creative, they do their own thing, and I've been lucky to have a family that doesn't restrict me or tell me what to do with my life. I think for that reason, I went to college because nobody was telling me to go to college. So I'm like, all right, I'll go. I wanted to get the heck out of Ajo. I grew up there for eighteen years of my life and it was all I knew. By the time I got to the city and lived in other places, it was a big culture shock. My entire undergrad experience, I was trying to figure out what I wanted to do with my life.

I was majoring in religion, political science, history. I ended up with a poli-sci degree just because I couldn't figure it out.

I went to college up in the Phoenix area, which is also O'odham land, Akimel O'odham territory, River People territory. Back in those days, the early 2000s, Sheriff Joe Arpaio was conducting a lot of raids in places where there was a boom in Yaqui villages, like the Town of Guadalupe, and it was awful. I recognized a correlation between the militarization we were experiencing back home, the Ajo border area, and these raids. But the city is in this bubble where it only recognizes itself. Back then, it was hard for people to see that correlation, that Indigenous rights are in the same conversation as what's happening with prevention through deterrence, militarization, and the detention centers. Since we, O'odham people, haven't been on people's radar, many don't think about these issues impacting Indigenous culture. But now that the border wall has crossed our traditional territory and Elbit-integrated fixed towers are on reservation land, we're a lot more in the spotlight.

For us, it's heartbreaking because it feels like a lot of the resistance is too late, especially since they've already built the wall across our sacred areas. It's been a horrible experience watching that and watching this ever-increasing militarization on our land. Sadly, my whole being is, I don't want to say defined by, but heavily influenced by this militarization that's happening around me. Every aspect of my life is impacted by militarization and by this encroachment by Border Patrol.

I graduated with my Bachelor of Arts in poli-sci in 2006 from Arizona State University, and that was when the raids started to get really intense. I briefly left to California, came back, and got involved in things like Copwatch. I remember a lot of traumatic things happening. Now, everybody talks about the screams and the cries of the children in the detention centers and the family separations. They said that Obama didn't make family separations, which is true to an extent.

There wasn't an intentional program within the detention centers to separate families. However, there very much were raids. And every single time there was a raid, there was a family separation. I remember the screams of those children. I remember the horror of it all, having to be there while these families are frantic and calling relatives because now all of a sudden a child is without care because their mother or father has just been taken by la migra. And why? Because Joe Arpaio and his posse are pulling over every person with a Radio Campesina sticker. They set up this stop in Guadalupe where they profiled every single resident. In Scottsdale, just a couple miles north, they wouldn't dream of profiling the residents. But here they were doing it in Guadalupe to everyone just because they're brown or of Yaqui descent.

In 2010, SB 1070 was making national headlines as the "papers please" law. That was a big turning point. Over in Phoenix there were the raids and Joe Arpaio. Over here in Tucson they had the ethnic studies fiasco, in which the school boards were taking away the teaching of history that didn't fit their white supremacist narrative. And my little town of Ajo was being targeted. Our school was singled out by school district administrators for not doing papers please with the kids. It had to do with kids from Lukeville or kids from Sonoyta taking the bus to Ajo schools. The right-wing media made a big deal about people's taxpayer money going to undocumented kids. Of course, they didn't use those words. They used words I don't like to use. No human being is illegal. But they would make these exposés, and they wouldn't tell the entire story. If they were telling the truth, they would acknowledge that Sonoyta is a historically O'odham village. Sonoyta, Son Oidag, is a farming community, and it's culturally tied with Ajo and the surrounding areas. The residents of Sonoyta often have family members in Ajo. They have papers. Tom Horne, who was the superintendent of the schools at the time, a very racist one, fined the schools. I believe it was $1.2 million. It's a lot of money for a small-town school that's poor. Since I went to that school, I know how poor it was. We had secondhand everything. We had books from the '70s with our parents' signatures in them. I used to talk about the contrast of going to such a poor school and these Border Patrol facilities with their fancy new vehicles and their fancy buildings. They have everything new and state-of-the-art compared to the kids in the same community who get next to nothing.

They wonder why we have problems. They wonder why there's substance abuse issues now. It's because this increased militarization takes our autonomy. It takes so much from who we are as people, and I think people, as colonized oppressed people, have unfortunate coping mechanisms. I've long made the argument that as substance abuse increases the more militarized enforcement gets, because that's what I've watched in my community. After 9/11, things got so militarized, and meth addiction skyrocketed and so did domestic violence. In my mind, that should demonstrate that militarization makes things worse, that it doesn't solve any problems. Treating children of a community as if they're criminals and not worthy of education creates this us-against-them scenario in which law enforcement—Border Patrol and police—is against the Indigenous people of the community and targeting undocumented peoples.

When it comes to O'odham, there's physical barriers like walls or mechanical barriers like towers that are constantly watching you. We have other issues to think about, like how do we connect with our O'odham relatives in Mexico who speak Spanish and are trying to learn O'odham, but because of colonizer policies in Mexico, no longer speak O'odham? We're facing that same thing on the north side of the border, but we have more language speakers on the north side. How do we get those language speakers connected with O'odham to the south of the border? How do we even communicate when most of us are speaking the colonizers' languages? We are trying to teach ourselves our language. We're taking all of these steps.

The wall construction began in August of 2019. From then until now, we've been infiltrated by construction workers. In December 2019, we had been completely defeated. I felt like my heart was on the ground, like they had won. They had taken over. Stopping the wall was not going to happen. So at the beginning of 2020 was the feeling of defeat already, before I even knew of COVID. In February, we had a ceremony at Quitobaquito, which was amazing. It was beautiful. We have pictures before they erected the border wall and we were able to connect with O'odham in Mexico. Then we just kept watching as they pumped out water from our sacred springs. It was a crazy time—we were dealing with all kinds of surveillance, infiltration, and lateral violence.

That infiltration continued during COVID-19. We already had these man camps taking over Ajo and our surrounding areas. For that to happen during COVID-19 has been scary and, as you can imagine, rage inducing. Workers are not taking COVID precautions in our hometowns and not caring because a lot of them are Trump supporters. You have these anti-mask people talking about their right to go to the grocery store without masks. Then you have a pandemic happening, and people are passing away. The spread was documented to have, at least in my hometown, started with the wall workers. It was inevitable, but it still pisses me off that they had no concern.

There was a big COVID breakout among wall workers, and so there was a big gap in time in which construction slowed down. Then we got word that construction increased, and it just took off. We had a lot of rallies, but we didn't do anything to physically stop border wall construction, which was disconcerting for a lot of us. I've been around a lot of badasses who held it down, and I couldn't understand why no action had been happening. But like I said, I felt really defeated.

October came around, and everything changed because we finally did something. Construction workers were going to be erecting the wall near the sacred site of A'al Wappia, Quitobaquito. Our thoughts were, we're just going to check on it and see where they're at with construction. We got to Quitobaquito early in the morning. It's a sacred place, so we had to do our prayers and our offerings. We also did a bit of checking up on the area. Since the area was making national news, it had been getting a lot of tourism that brought a lot of unwanted environmental effects.

As we were walking around the springs documenting, in separate instances, my friend and I heard drilling and construction. I remember becoming really frantic. I have this home video that I haven't released, but it reminds me of *The Blair Witch Project* because I'm running around frantic, breathing hard, and you can hear the workers drilling in the background. I'm screaming for my comrade. I found her finally, and we both looked at each other, and it was like we had come to the same conclusion. No, not here. Not now. We ran towards the construction and got in the way of it. We didn't plan it. It just happened. It felt driven by that morning, by that prayer and reflection.

In retrospect, it was a little insane because we were out there by ourselves. My cell phone doesn't work out there, but my friend was able to connect and get in touch with some comrades. We had about an hour in which we were holding the construction off by ourselves, and then the media got there and was able to get the footage. Immediately, once the camera was on Border Patrol, their behavior changed and they changed their plan completely. Border Patrol took off and had to reevaluate and come back. We got arrested, my comrade and I. We were lucky that it got filmed so our jail time wasn't in vain. We ended up staying at Core Civic for two days, not knowing how long we were going to be in there. We had no idea that our comrades were going back to the construction site. They were holding off construction workers. That was another badass moment in which I am proud of my people. After that, we regularly had standoffs with construction workers and Border Patrol and Park Service rangers. There's footage that shows Border Patrol and Park Service officers being violent with water protectors.

It's been intense going through all that during COVID. We always had that in mind and remained masked up. Organizing during COVID is a whole different thing. Having that obligation to stand up for the land while a pandemic is going on is so much. It's been a crazy year—it's been a crazy two years, three years. Before the wall and before Trump, it was the integrated fixed towers. Now, it's everything all together: it's the towers, it's the wall, and it's COVID. And it's insanity—to sum up 2020.

I would like to believe that the Biden administration is feeling enough pressure that they are actually going to do something about the wall, but in having some familiarity with Democrats and how they've used the right wing to cover up their past stances of supporting militarization, I'm not so hopeful. We got the Border Patrol Center in 2010 during the Obama administration, so most of the infrastructure that they've been using to carry out the building of the wall wouldn't have been possible if it weren't for the Obama-Biden administration. Do I have hope that they themselves, under their own initiatives, will take down the wall? No, I do not. The recent history of not only militarization, but Border Patrol, deportation, and pretty much the whole shebang is what we're fighting. What I think is different now, what I have hope in, is pressure.

The horrible experience that we've had under Trump has gotten a lot of people's attention to these atrocious policies. Prevention through deterrence has been happening since 1994. Since Trump has been so bad, people are now seeing the policy for what it is. Now Democrats won't have such an easy time fully supporting militarization efforts. It's up to us to keep that pressure on Biden. There are many Democrats between the Senate and the House that will get away with supporting funding for walls if we don't keep them accountable. They've already done it. The COVID relief bill has a ton of money for the wall and there's not a lot of talk. If the Democrats were functioning in the way that they are supposed to, you would see every piece of legislation that has funding towards the wall being called out and every person who voted for it being called out. But you're not seeing that. That makes me really cynical and disappointed. Going forward, I feel that the task is upon us to make sure that we keep these issues being talked about. That's my biggest fear, that once the Biden administration comes to power, people will live in the same ignorance that they did before and not pay attention.

Even though we didn't stop the wall, all of the direct action gave me hope. We're being that example for the future generations. We're showing them that we did do something. It wasn't just standing there with a sign. We had our own O'odham shot by rubber bullets. We're standing up and our children are seeing this, and they're going to be warriors because we stood up and we were warriors. That feels really good.

I had almost completely lost hope in my heart, in my being, in everything. I had been depressed and feeling really low. But coming together and standing up revitalized hope that we can take back our land and educate people on who we are. We're still here, and we've let them know that the more they want to militarize us, it's not going to be easy for them. I want them to know that this is not just like *cha ching*, they're not just going to make money. We're going to make it difficult.

For me, it's not just about the direct action, it's about the whole package. It's about revitalizing our languages, connecting with our relatives on both sides, not letting those borders and barriers come between us while doing what we can to physically dismantle them. These walls do need to come down.

Envisioning the world we want to live in is something that we think about a lot as land and water protectors. We have to keep in mind that in our battles, we're not going to see direct results tomorrow. For example, we didn't stop the wall. Unfortunately, in our sacred areas like Quitobaquito, they've already built a border barrier. It's heart-breaking to see that before our eyes. Sometimes we can take it as a loss. However, we are thinking about the future generations. It's helped me a lot to think of it that way. We may not see the end of capitalism in our lifetime. We may not see justice for Indigenous peoples in our lifetime. We may see some pretty horrendous things to come, but we need to lay the groundwork for those future generations so that they know that they're supported, that their grandmas fought and will continue to fight for them. Laying the groundwork for this setting in which the battle can be won in the future, I think that's what it's about. It's also about ingraining good values, changing the dynamics of the colonizer frameworks, and breaking down those borders and barriers that are between us literally and figuratively.

We have our tradition to hold on to. I remember visiting the spring and feeling that loss, because we had just seen how close the wall was to this spring. But I also felt a sense of rejuvenation, like our culture is still here, we're still collecting and harvesting our fruit. We just had the Gregorian New Year, but our new year as O'odham is in the summer. It's with the seasons changing and the Baidaj harvest. That is when the saguaro cactus, our Ha:san, has this wonderfully amazing-tasting fruit, and that's one of the things that I have come to love and get excited about every single year, the harvest of the saguaro fruit. We're going to exist on this land, and just look at how beautiful it is, you know? Self-restoration is a combination of participating in our culture and experiencing the new year and the monsoon rains that come with the new year. We're doing what we can to decolonize. We try to literally do that, and what better way than to learn our own traditional foods, our own traditional medicines.

During COVID, as horrible as it has been, we've been exchanging recipes on breathing remedies and our traditional ways of healing. A lot of that has been rejuvenating to my soul and to my whole being in every sense of the word. I love our culture, our spirituality, our stories. That has helped me get through these trying times.

Nellie Jo David works to strengthen Indigenous rights and autonomy on the imposed US/Mexico borderlands intersecting the Tohono O'odham Nation. Cofounder of the O'odham Anti Border Collective, a grassroots group dedicated to maintaining connections despite colonial barriers, she is from Ajo, Arizona, traditionally Hia-Ced O'odham territory, just west of the Tohono O'odham Reservation, North of Mexico. Nellie obtained her JD with a certificate in Indigenous law and policy from Michigan State University in 2014. She is currently working on her SJD at the University of Arizona in the Indigenous Peoples Law and Policy Program.

PRAISE POEM FOR ALL THE HOMES I'VE STRUNG TOGETHER

BY KINSALE HUESTON

I am unsure of how many times I let "home" slip
into praise poems during these months. Praise heartbreak
in Connecticut, my learned home a city that ignites:
Her children topple statues of murderers and I think of this
when the absence that has taken root in me dreams
of winters. Bitter wind is still better than stale air,
so I imagine peaches passed around in paper cups
and piss-poor poetry scrawled on blackboards; my walks alone
through archways we have not yet torn down.

In a second language, I praise our mothers and their migrations.
When the world was blistering, they pressed food to mouths,
strapped boxes to truck beds. I don't mean the mothers by blood—
I mean the relatives who braided homes together again, for whom
I write praise poems about existing: mornings with the dogs, with our
hands
in cool water, teaching the children how to turn the soil.
The sound of something cooking where once
there was nothing. Soft breathing into a stretch of morning sunlight,
while outside the willows murmur in their new solitude. Every patch
of earth holds the people; the people hold each other.
Praise the slow days like this. Our mothering of each other,
Our giving.

In a third space I create, I must find home or bend into
the aching. I write of dark bellies and a stranger
who teaches me to pray. It's much too easy to believe

there is an answer, somewhere, to the body
I feel like I am lacking. I don't praise the empty, but maybe
the familiar that can fill it. Once, my grandmother
was lonely, too. From her,
a home for my fossil. I, once smaller
than a blue corn kernel, must then have come
from lonely. I praise the home we've built
from that, too.

Biography on page 132.

STEEPED IN THE TEACHINGS OF INDIGENOUS MATRIARCHY

BY AHJANI YEPA

As told to Alastair Lee Bitsóí and Brooke Larsen
January 2021

He` Teh No-ah, I am Ahjani Yepa, I am Jemez Pueblo and Anishinaabe. I am Badger and Crane clans. I was raised by my mother and her family including my aunties, grandmothers, and many matriarchs in Jemez Pueblo. I am incredibly grateful to have been blessed with Pueblo values and Towa language. From an early age I learned the importance of listening. I learned how to sit still in the plaza and how to dance during our feast days.

I was also given teachings from my father in the summers, learning how to sit still in city council meetings, how to march, and how to speak up for Indigenous rights as I watched him organize in the '90s in the Bay Area. I am doing my best to live the values both my parents taught me.

I started speaking out myself in 2009, while I was a student in Grand Rapids, Michigan, against new proposed coal infrastructure. In 2016, I answered the call to join the resistance against the Dakota Access Pipeline. I was part of the Pueblo camp at Océti Sakówin. In those moments we did our best to live the values we were each taught from our Pueblo homes and communities. I helped to feed and care for my camp family; together we experienced so many miracles and also a lot of unjust suffering inflicted by the hands of the US government, not much different than every generation before ours.

After returning home from Standing Rock with people who became family, we built solidarity together across Pueblo and Diné people to protect our communities and sacred places like Chaco Canyon from fracking and extraction. When I started learning more about the protection of ancestral landscapes and our deep histories as Pueblo people, I realized that our homelands do not end at our reservation borders. When I first visited our ancestral home sites and connected with petroglyphs as far away as southern Utah, it was like remembering something I had stored deep inside myself. Sharing that experience with as many other Pueblo people as I can has been part of my healing process and journey.

I see a direct connection between our struggles as Indigenous people and all oppressed people because we are all colonized and displaced people within the United States. As an Indigenous person, I am continuing in the tradition of generations of Indigenous people before me defending land and protecting water. We have survived despite era after era of federal Indian policy meant to eliminate us and the violent dispossession and removal from our homelands. The lands that are vital to us as Indigenous people are often not within reservation borders or under tribal stewardship. Indigenous people were displaced and forcibly removed from lands that are now known as public lands under the Department of Interior, such as national parks. Private ownership of land is also a result of the country's violent history. Justice is the healing we need for the historical and generational trauma. Capitalism, colonialism, and white supremacy are at the root of the issues we collectively face. These violent systems have impacted not just Native American people, but all people globally.

Reflections
Fractal projections
Fragmented versions
of self
Shifting time and space

Finding a place
When spirit is free
Reflected
in sand and earth
The land she called me home
Falling victim
To boundaries and boxes
The sign that reads
No trespassing
It's a game we play
Just a trap in our head
It's the land and it wants us back
She wants us back
We want her back
And the only thing stopping
Are the lines in our minds and
The boundaries and borders
Tortured and murdered
Onto the land

During the pandemic, it's been surreal witnessing how Indigenous communities are suffering. The pandemic has been a magnifying glass showing all of the cracks, all of the inequities that have been here not only in Indian Country but across the US. I'm experiencing the same systemic problems of each generation before me. I am thankful, at least, that finally we are looking directly at what's been wrong. This year has been a deep, reflective time for the United States. I'm grateful for that. We are opening up a lot of people's eyes to the reality that this country is in. This nation hasn't ever been equal.

I've been reflecting about how abruptly everything stopped, and how necessary and needed that was for us to slow down as humanity and as people to give our earth a break, to reassess our personal lives. We all learned how to wait again. I remembered the value in taking my time, the value in meeting with others face to face, how simple that is, and how much I had taken that for granted.

For Indigenous communities, this moment has opened up new possibilities in the face of the obstacles we are surviving. We are seeing the first Indigenous woman to hold a cabinet position. To see real prayers manifesting is beautiful. I have a lot of hope, and it's absolutely coupled with fear. How to hold all that's happening in the world and maintain an outlook for the future is something I have been digesting all year.

We have been practicing
our personal ceremonies
taking care of each other
holding prayers for future blessings
Emerging from waters
warmed from her depth
blessed and renewed

I have been so lucky to have kept my child home and in good health this year. I'm sure we are facing the challenges that all kids and parents are having while navigating this pandemic, but what I am most grateful for is the ability to have my child home with me. We have taken the time together to reteach and reinforce Indigenous language, concepts, and values from home. When I think about generational trauma within the context of boarding schools, we now have a chance to educate our own children from home, something that had felt privileged to non-Natives before the pandemic. Our family chose to unschool and decolonize our learning this year. My hope for her is that her world is more loving and less painful than the one our grandmothers lived in. I hope she holds the lessons and teachings we shared this year close to her when she chooses to return to school and we all venture back out into the world.

Daughter: 'Is this where our ancestors came from?'
Mother: 'Yeah.'
Then she asked, 'Where did they go?'
I said, 'Well, they went to Mesa Verde, Chaco,
eventually they came down to Jemez,
and to where we are now.'
And she said, 'So basically, we're them?'
'Yep, basically.'

We need to steep ourselves in Indigenous teachings. Indigenous knowledge is the solution to the climate crisis, and right now we need to allow for young Indigenous people to have access to our elders and to learn what teachings we can. Our job will be to interpret how we apply them now and in the future. That's always been the job of each generation, and for this generation the weight of that has never been heavier. Our personal conversations and how we share knowledge with each other as peers is an example of how nonlinear knowledge transfer has become. One of my strongest and most valued teachers and mentors is my younger cousin; I love and respect her as I would my elder. More and more now, I'm finding my teachers are my peers, or even people that are younger than me, like my daughter. She teaches me things every day.

I was told that when we are gifted with a teaching, we have a responsibility to share that teaching. I'm working to share my knowledge with my daughter. I teach her to understand that responsibility. I believe that she will carry and pass that on. I pray we continue to remain connected with our land, our cultures, our traditions. Access to our ancestral land is imperative to transferring our ancient knowledge, and practicing of Indigenous knowledge by Indigenous people is vital to maintaining balance in the face of climate change. Our teachings are essential for regeneration of our earth, and our people are essential. We move in cycles: time and the seasons, our moons, each passing day, month, and year. As people of the land, we are Indigenous everywhere we go—we have remained connected. With the closing of old cycles and opening of new ones, I pray that we open the cycle of healing and justice for all oppressed people.

Ahjani Yepa (she/her) is a Hemish/Anishinaabe mother and Indigenous feminist organizer dedicated to the protection of Indigenous Peoples and rematriation of ancestral places. She is cofounder of the Women of Bears Ears, and a co-author of their New York Times *op-ed: "Women of Bears Ears Are Asking For Help to Save It."*

CLIMATE JUSTICE MEANS PRISON ABOLITION

BY UYEN HOANG

Since 2019, I have been a volunteer with the University of Utah Prison Education Project (UPEP), a program that provides on-site college courses to incarcerated students. As one of the teaching and learning assistants (TLAs) for the program, I helped facilitate college courses on gender studies. Each week for three hours, the TLAs and students would sit together to discuss and question society's unforgiving norms. How could we envision a future where harmful power structures, such as gender roles, no longer exist? How could we imagine a world where long-established practices and roles are obsolete? We'd find ourselves talking in circles, unsure of how to undo or fix systems deeply entrenched in upholding oppression. It was nights like these that led me to learn more about abolitionist work.

In a nutshell, abolition means community care. Abolitionists work toward making prisons and police obsolete while implementing preventative care to resolve inequities and ensure that every person is able to have their needs met. It is transformative, imaginative, and involves radical love. Abolition also requires uprooting harmful systems such as capitalism, oppression, imperialism, colonialism, and the patriarchy, which are also the very same things that perpetuate the climate crisis.

When I was an undergrad in Environmental and Sustainability Studies, many people were confused when I told them I volunteer with UPEP.

"Why are you involved with something that doesn't relate to your major?" they'd ask.

What they did not realize is that environmental justice also means prison abolition.

An investigation from the Public Accountability Initiative released in the summer of 2020 found that many of America's largest oil and gas companies and private utilities, along with the banks behind them, are funding police foundations throughout the nation. In Utah, Chevron, Dominion Energy, Rocky Mountain Power, the Church of Jesus Christ of Latter-day Saints, and Wells Fargo are among the largest donors to the Salt Lake City Police Foundation.

This report came out as the COVID-19 pandemic and mass uprisings for racial justice revealed inequities in health, environmental hazards, and policing.

Climate change is causing an increase in dangerous disease outbreaks. The planet's natural defenses against newly emerging infectious diseases are weakening and prone to extinction. Deforestation, loss of biodiversity, and conversion of natural lands to economic development are all contributors to the spread of disease. Like all aspects of the climate crisis, the projected increase in pandemics will impact marginalized communities at a disproportionate rate. We have seen this in prisons throughout the COVID-19 pandemic.

A study published by *JAMA: The Journal of the American Medical Association* shows that incarcerated folks are 550 percent more likely to contract COVID-19 and are 300 percent more likely to die from it than the general population. Prisons lack regulation and appropriate safeguards against environmental and health hazards. Pandemic safety recommendations from the CDC are difficult to enforce in prisons. An incarcerated friend whom I correspond with has reported that the Utah prisons did not even quarantine those infected, in an effort to expedite herd immunity. The only accessible medication to treat the novel virus was aspirin, and most prisons lacked adequate testing and personal protective equipment (PPE).

Communities of color, which are disproportionately impacted by pollution and police violence, have also been disproportionately affected by COVID-19. When people of color seek health care, they are met with more racism. Susan Moore, a Black medical doctor, was hospitalized due to COVID-19 in December 2020. Her white doctor denied her pain relievers and minimized her condition. She believes she was discharged too early, and in a viral video she states, "I was crushed… He made me feel like I was a drug addict, and he knew I was a physician." Her treatment was later corrected; however, she passed away from COVID-19 complications two weeks later.

Stories like Susan Moore's are not uncommon. When people of color seek help or refuge from institutions that are meant to provide aid, they often receive the opposite. Systems and institutions—such as corporate health care, capitalistic monopolies, prisons, and police—that are built by predominantly white, cisgender, heterosexual, wealthy men only benefit those within the same demographic. Early police departments, staffed by white men, were created to preserve chattel slavery in the eighteenth and nineteenth centuries. To this day, 65.5 percent of police officers are white, according to the Census Bureau. Policing is just one of the structural ways white supremacy has sought to protect the status quo.

Black, brown, and Indigenous communities are more vulnerable to pollution, largely due to racial segregation set in place by government policies and economic factors. As a result of this systemic racism, communities of color have been continuously disempowered politically and financially. Neighborhoods of mostly non-white people have lower property values. This allows the fossil fuel industry and corporate entities to more easily acquire nearby lands to excavate, extract, and exploit.

A stark and unjust cycle of oppression revolves around corporate polluters, a history of systemic racism, and policing. Ruth Wilson Gilmore, a prison abolitionist and professor at the City University of New York, sums this up by stating, “The places where inequalities are the deepest, the use of prison and punishment is the greatest.”

Law enforcement disproportionately polices Black and brown neighborhoods. In 2020, Utah police shot at a record thirty people. Of those fired at, seventeen died. The youngest shot at was thirteen years old. Forty percent of the shootings occurred when law enforcement was responding to a call regarding a mental health crisis. How can a profession that stemmed from slave patrols enforce justice and protect others? Since their inception, police have been a force of violence against Black people and the impoverished. It’s as if it is their job to brutalize others…because it is precisely that.

When frontline communities organize for equal access to opportunities, clean air, water, land, and justice for their people, they are met with more violence and criminalized by the state. The militarized response to the water protectors at Standing Rock in 2016 serves as a blatant example. Water protectors mobilized against the completion of the Dakota Access Pipeline, which would transport 470,000 barrels of crude oil through sacred Sioux land and threaten to contaminate their entire water supply. As water protectors sought to defend the land and their quality of life, they were met with water cannons, tear gas, rubber bullets, bean bag rounds, pepper spray, Tasers, and sound weapons. Police stood in full riot gear while water protectors gathered in prayer, and the police intentionally released inaccurate information, demonizing the water protectors fighting for their water and land rights. The pipeline was completed in 2017 and leaked at least five times the same year. Since then, over a dozen states have tried to criminalize pipeline protests.

In Salt Lake City, protestors are facing felony charges for fighting against the proposed Inland Port. The port will turn seven thousand acres of land near the Great Salt Lake into a warehousing facility and freight distribution center. The Utah State Prison will also be relocated to Salt Lake's western quadrant on previously undeveloped land. The new prison will be laying down foundational infrastructure that will allow the nearby Inland Port to more easily be constructed. The ground it is being built on, primarily wetlands and liquefied soils, is vulnerable to groundwater contamination, flooding, and acute earthquake damage. Thus, those incarcerated at the new prison will be more at risk of contaminated or toxic water, as well as injuries or death if an earthquake were to occur.

Once incarcerated, people previously impacted by environmental harms remain the most vulnerable to disease and the climate crisis. Prisons lack proper heating and air conditioning to keep incarcerated people safe in extreme heat waves and winter storms. Incarcerated folks are also deprived of quality and nutritious food, warm clothing, privacy, health care, clean air, clean water, proper tools towards healing, and the list goes on. When someone's basic needs are not met, they become more susceptible to disease and mental illness. When a climate catastrophe hits, incarcerated people have scarce protections and resources. In 2019, officials refused to evacuate the Ridgeland Correctional Institution in South Carolina despite its location in an evacuation zone. When Hurricane Florence arrived the year prior, three prisons had also refused to evacuate the incarcerated.

Climate change has forced more than 79.5 million people to flee their homes. As climate refugees are displaced due to flooding, storms, desertification, droughts, and other climate disasters, countries have been unable to provide adequate and timely assistance. As many migrate to different countries, instead of being met with open arms, they are met with cruel, militarized responses. Families who have been forced to flee are being separated from one another and detained in immigration detention centers, often owned by private prison companies, with inhumane conditions. From toxic prisons to migrant cages, incarceration has only ever exacerbated suffering.

This cycle of environmental destruction, brutalization, and criminalization will inevitably continue unless institutions that police and pollute our communities are abolished. If combating climate change is about eradicating fossil fuel emissions and fighting for equity, then we must also eradicate the carceral state that upholds the fossil fuel industry and further perpetuates the degradation of our planet and the suffering of all living beings.

But abolition is also about birthing new systems. As we defund the police, prisons, and fossil fuels, we must reinvest in systems of care and regeneration led and owned by impacted communities. This includes education, mental health, nourishing food, renewable energy, social workers, green spaces, clean water, and growing systems of accountability.

In order to dismantle racial capitalism, white supremacy, mass incarceration, and patriarchal domination, we must begin to center the voices, experiences, and needs of the communities who have been historically underrepresented and ostracized. This all begins with imagining a new world.

How do we build a world that is accessible?

How do we build a world that makes us feel safe?

How do we build a world that makes us feel supported enough to be our best self?

How do we build a world where we can trust one another to build solidarity?

How do we build a world that does not attribute our worth to our productivity?

How do we build a world that is free from borders and free from land ownership?

How do we build a world without a need for Instagram's "antiracist" infographics?

Your turn. How do we build a world that is _____?

There was a point in time at UPEP when the students began to ask "what if" less and began taking action themselves in creating a world they wanted to see. One of the students started a feminist literature book club; another student was devout in correcting his peers' gender-derogative language; and another student began taking steps to raise his grandchildren in gender-creative ways. Dreaming provoked the "what ifs" and collective care inspired the rest.

In the past, a world without slavery or without segregation was unimaginable to many. While these issues have not been completely eradicated, abolitionists who dared to dream helped improve and move beyond these systems.

In dreaming, we must contemplate our own role. Whether it be to disrupt, resist, strategize, heal, share knowledge, tell stories, give care, guide, build community, or find joy, we must continually center collective thriving.

Let's imagine and create a world that values people and planet, not prisons and pipelines.

Uyen Hoang is a Vietnamese-American environmental justice organizer based in the occupied Eastern Shoshone and Goshute lands, also known as Salt Lake City, Utah. Her background is in ecojustice education, studying how misogyny and climate denialism intersect and how to combat it. She spends her free time imagining anticapitalist futures, chasing new menus, and petting her cat, Soju.

PART 3 FULL MOON

During the Full Moon, the moon is most illuminated. In this section, contributors shine light on the solutions and new stories being birthed. They dive deeper into what a regenerative and just future looks like using imagination, reflections on current praxis, personal narratives, and plans for structural change.

WATER GODS OF THE NEXT WORLD

BY LINDA HOGAN

This is a time when I wonder if we did truly enter the fourth world
as they say, or if that time is yet to come, because we entered this
world
for peace.
I ask the man whose teachers are clouds; they live in caves and
mountains.
They live in thick forests and some of them come out
through broken windows
of houses made with round stones. Others come from the crashing
waters
where seas converge, the Pacific and Atlantic with the Southern.
Cloud teachers and water sages wear white. Sometimes they live
with the world of birds. When they pass over this world, they see
suffering
and they weep rain.

The good man tells me, If you climb the peaks, sometimes you hear
them singing.
Clouds are people I don't know except they are mostly humble
crossing our sky. Still, I want to know if we have yet another realm
to enter,
to emerge from our ways of living now, to climb a reed, or leave our
crystal cave.
I thought the goddesses of clouds and gods of rain
were merely the breath of wind, but some are the story of storms
held back too long from telling they've seen how hard the world has
grown.

In the beginning, before there were flower people, before bees,
were birds and sky.
There was no land. Birds rested on water held by clouds.
For others, a miraculous spirit was born in the presence of animals,
like the rest of the poor with an earth but no clean water to drink.
Tree Woman arrived. She planted and rooted and changed the course of water
so life could go on without illness. For this they were grateful.

These are only a few of the stories and songs that exist. All are true.
In the dry land corn waits for the holy presence of clouds
and human beings dance barefoot on hot earth with their love
for rustling corn.
Others live in the place where waters meet,
where the people climb mountains
with the deer to watch those clouds fly in, dark, full, ready to give birth,
and soon the water breaks and it flows.
Then those of the water clan know the songs that call water forth.
They place tobacco prayers on surging rivers where clouds rise
and see their prayers float on the water, then carry the songs
to greater waters.

Yet how I do wonder if one day we'll climb from this world
into the one of peace where countries do not harm their own,
where rumors do not end in war, leaders have wisdom and integrity,
caring for the free clean water of this world.
They know the stories of all people are true and respect them.
This is how I know we are not yet at the beginning of the world
as some of us were told.

Biography on page 12.

WATER IS A TEACHER
LESSONS FOR A JUST TRANSITION

BY NICOLE HORSEHERDER

As told to Brooke Larsen
May 2021

Nicole Horseherder founded Tó Nizhóní Ání (TNA), which means "Sacred Water Speaks" in the Diné language, in 2001 in response to Peabody Coal Company's excessive use and waste of the only potable water source the Diné people have on Black Mesa.

Peabody began mining coal on Diné and Hopi lands on Black Mesa in the late 1960s. Coal mined on Black Mesa was then used to power the Navajo Generating Station (NGS), which opened in 1974. NGS was built to power pumps that moved water from the Colorado River to Tucson and Phoenix through a canal called the Central Arizona Project. Electricity generated from NGS was also sent to cities across Arizona, Nevada, and California. While NGS supplied water and power to Southwest metropolises, many homes on Black Mesa still don't have electricity or running water. NGS was the largest coal-fired power plant in the Western United States and the largest source of nitrogen dioxide emissions in the country. NGS and coal mining on Black Mesa stopped in 2019 as costs of operation became prohibitive compared to cheaper renewables. The power plant was officially demolished in December 2020.

Mining on Black Mesa stripped people of their homes and sacred places while depleting and contaminating the water. Pollution from NGS also severely impacted the health of Diné and Hopi people. However, the mine and power plant also created hundreds of jobs and significant revenue for the Navajo and Hopi Nations. Nicole Horseherder works to address this tension through a just transition framework.

Years before mining started on Black Mesa, the federal government began an elaborate campaign to forcibly reorganize the Diné and Hopi traditional governing structures and put in place governments and individuals who would favor extractive industries. The government reorganization in the Navajo and Hopi Nations served as the model for the 1934 Indian Reorganization Act. Land disputes between the Diné and Hopi people were created by the federal government, and eventually over thirteen thousand Diné people were forced to relocate so that mining operations could expand. This is a long and complex story of the relationship between settler colonialism and extractive industries that we encourage readers to research.

I work for the organization Tó Nizhóní Ání (TNA). I helped cofound TNA back in 2001. It's an organization that's based on Black Mesa, and its original purpose was to protect the water sources of Black Mesa, especially to protect the water sources from the coal industry. We realized that if industry was going to continue using the enormous amount of groundwater that they were using, there would be water shortages and changes in water levels. Our entire hydrologic system of Black Mesa would be compromised. We didn't know what that would look like, but we always wanted to give the benefit of the doubt to the water. There were just too many questions. For all the experts, all the hydrologists, and the number of studies that have been done on the aquifer, nobody could say for sure what the true impact of both coal mining and groundwater mining was going to be on Black Mesa. So that's how we started, and today the water still guides us. That's led us to mining and compliance issues, pollution issues, and other environmental justice, social justice, and economic justice issues.

In 2018, TNA, along with other Indigenous organizations, helped block entities, including the tribal enterprise Navajo Transitional Energy Company (NTEC), from acquiring the Navajo Generating Station. These corporations thought that they could step in and acquire the NGS plant, keep it running, and get some kind of profit from it. We just said, "No, it's gonna shut down when the contract ends [with Salt River Project], and it has to." At that crossroad [with NTEC], the non-Navajo and the non-Indigenous allies stepped away. They basically said, "At this point, this is a tribal issue. We can try to support you however we can, but we've gone as far as we can go with you and you've got to take it forward." The fight to finalize that closure rested on Indigenous organizations, and Tó Nizhóní Ání was there leading this effort to make sure that NGS shut down for good and didn't end up in the hands of somebody else.

Our work has expanded. From there it just became about transition, and what a welcome moment that was. I am so tired of fighting coal and power plants. I'm tired of hearing how reliable that energy is and how important it is. "Well, you drive a truck, don't you? How do you expect to get gas? How do you expect to get around?" I'm tired of those questions. I'm tired of people ignoring the environmental impacts of the fossil fuel industry. I'm glad it's done and going away. I'm really happy, actually, to be working on transition issues. I'm a little bit more relaxed. I'm have more time to take care of myself and the people that are important to me. I didn't really have the opportunity to do that when we were fighting to get the coal mine off of the aquifer, trying to shut the slurry line down, and then going after the coal plants. For that industry to continue operating meant the continued use of the water. And that was not going to happen.

In December 2020, NGS was demolished. To be honest, I was not going to go to the power plant on the demolition day. I had just hired a media organizer, and his first job was to go out and do the live stream and provide that information. He's the one who said, "Do you want to go with me and help me hold my equipment?" I went, and then of course I dragged other people along with me. I dragged Marshall, my husband, along.

Because of how long we've been doing this and how intense it is, it's hard to be moved by something like the demolition. For us, it's always, "What's next and what's next after that?" For me, the win was twenty months earlier when NTEC put out a press release and said, "We're not going to pursue NGS anymore." Then in a matter of weeks, Peabody put out their statement saying that they are going to shut down the mine. That, for me, was the win and biggest relief. That was hope that this work that we're doing is worthwhile and that it's making an impact in a good way for the people.

Watching the demolition of the smokestacks was kind of like the aftermath. The part that was exciting was the number of people watching the live stream and commenting. The type of comments that were coming in were "Yay, it's gone, now we can see the skyline again, hooray for the environment," and then "I'm so sad, this power plant put five kids through college." They were both important comments, because all of it is true.

I always have in mind the people with jobs at both the mine and the plant and the revenues that the Navajo Nation depends on so much. I always keep that in mind. It's the reality.

However, the bigger picture is important. The bigger picture is why we do what we do. For instance, if you're at a Navajo Nation Council session, and the closure of NGS is the topic, fifty mineworkers get bused in along with another fifty plant workers, and then you have these grassroots people outside. Just listen to what each is advocating for. One is advocating for their jobs, so it's a me kind of thing—me, my household, and the benefits that the plant or this job has on a personal level. Then if you listen to the, for lack of a better term, environmentalists, or, better yet, the people who live off the land on Black Mesa and are not benefiting from the payroll, their concern is the decreased levels of water and the disappearance of springs. In addition, they might be concerned about the impacts to the environment, the vegetation, the pollution, having to be moved to make way for mining, living with the noise.

If we lose our water, the entire plateau has to go somewhere else to get their water. I keep the mine and the revenues in perspective. But without water on the plateau, there is no life. Period. There is no life. It is something that affects so many more people and so many more lives than any of the jobs at the mine or the plant together.

People talk about the economic impacts, and this last fight to close NGS, we had the economics on our side [because renewable energy is now cheaper than coal]. Previously when we've fought the utilities, the economics were not on our side. For industry, the bottom line is all about money. It's not about finding the best solution. It certainly isn't about conserving or putting in protective measures for the water. It is about continuing the operation in the least costly way for the operator and who can we continue to exploit so that profits are made. That's what it's about.

When you look at the whole situation, there's nothing that can replace the water, the environment, the air—these are the elements of life. There isn't a way to justify the jobs and the revenue over the environment. But I've heard every argument for the continuation of this type of industry. None have acknowledged the impacts to the people and the environment. When people say coal is more reliable than solar, I don't give a shit. Nothing is good about jobs that kill. If we have to move to an energy source that is considered unreliable yet it is clean and cheaper, then I can wake up with the sun and go to sleep with the sun. In other words, I'm willing to make the sacrifices for clean renewable energy. If it's going to hurt somebody else, I don't need to have my lights on at night. It's a matter of being more conscious about yourself, your life, the impacts you have, and the impacts the energy you use has on others. That's the perspective I come from, and I attribute that to the water. The traditional teachings of that element of life and the way of life that is still vibrant on Black Mesa. Connection with elders and their teachings enhances one's understanding of what is most important in life. Having gone to school and come back, my Western training helps me understand the technical and legal information and translate it to my community members and then translate the traditional wisdom and teachings back to both tribal and state leadership.

Coal mining depleted the water on Black Mesa, and it has also contributed to the climate crisis, which is causing drought across the Southwest. The climate crisis has been an ever-increasing problem. It is always part of the bigger picture, and a bigger problem that needs more discussion. It is a problem that can only be solved by the parties that brought these coal plants to the Southwest. As long as they control the utilities, the transmission lines, and the politics, they have to get serious now and build clean and renewable energy and prioritize communities that have been reliant on coal plants for decades. In the last twenty years that I have been doing this work, we've never been able to get to the critical discussion of climate change. I've never been able to talk to the Navajo Nation or Resource and Development Committee of the Navajo Nation Council about the impacts of climate change to our Navajo communities. We have never had a discussion about how coal-fired power plants and coal mines on our lands are having a tremendous impact on our Indigenous brothers and sisters in the Arctic, how they're literally losing their habitat and ability to hunt for their food. The wildlife are also losing their habitat. What I find most shameful is when Navajo leadership talk about how much we love our coal-fired power plants. I just find that ironic.

The impacts of climate change are before me every single day. I haven't had a harvest since 2016. That was the last time I had a full acre and a half of corn. That year we made so much steamed corn and kneel down bread. We dried corn, ground some, and stored about half of the crop. Since then, I haven't had a successful harvest because it's so dry. We just don't have enough precipitation. Our monsoons, which we rely on to bring our corn to harvest, are not predictable anymore and haven't brought enough rain. So we're losing our ability to feed ourselves. There's no corn that any grocery store sells that can replace our Navajo corn, especially the corn that grows on Black Mesa, which is different from, say, Hopi corn. If you don't know the difference or you're not aware, then they'll pretty much look the same, taste the same. But the corn on Black Mesa that's grown higher up has a shorter harvest than the corn that's lower down at Hopi. Hopi corn has a slightly longer harvest by a couple of weeks or so. That's how different it is, and we're just less than thirty miles apart from each other.

Climate change is real. It's here. We're dealing with it. For Indigenous people who live off the land and rely on the seasons and the reliability of the monsoons, we're losing our ability to feed ourselves right before our eyes.

To address the climate crisis, the greatest thing that needs to change is our mindset. We are good at lying to ourselves. We have to change our priorities and what we value. It seems like that would be easy, but for human beings, those are the hardest things to change, especially if you're used to a certain way of life. People don't want to give up comforts. But the people who know the crisis is here are the people who spend time outside with animals and with gardens.

The thing that keeps me going is the children. They give me hope. When you have been doing this work that I've been in for so long, you understand simple truths like, if I woke up today, I should be thankful because waking up is not guaranteed. What are we supposed to do as adults, especially as parents? Should we roll over and let climate change take its course? I don't know what other parents are doing, but I'm not doing that. A sense of obligation to the children keeps me going. There's nothing that can replace that energy and perspective of young people. They have so much going for them. We put them here, this is our doing. Why wouldn't we try to do something while we can so that they can actually live the life that we gave them?

Moving forward with a just transition means a lot of things to so many people. I often talk about just transition as happening at different levels. The most basic or grassroots transition work happening is that of food sovereignty—learning to grow your own food (again) and learning to grow food that is appropriate for the region or climate where you live. Not trying to grow cotton in the desert, like we do here in Arizona, but rather get back to the native corn, which is resilient to drought, hardy, and has a shorter harvest period. That's the most grassroots-level transition that anybody can do. Because of the drought, it's often more feasible to do backyard gardening where you can manage the water and shade better. Another project could be replacing your power with solar power by adding solar arrays to your house.

Another level of transition is when collectives are formed. For example, a community or a housing complex could get together and decide to invest in a small solar farm to power their housing complex. They manage it and maintain it and get all their power from their solar project. Similarly organized community members could also bring their crops together and have farmers markets. Farmers markets provide local crops to local people and cut out the large carbon footprint of grocery stores and chains. This would be the next level up from what people can do as individuals.

What I work on is a few levels higher than that. I am working at the commercial-scale development of solar energy projects. I'm working at that level because in these types of projects that bring back old partners and encourage new agreements, assets left behind by the coal industry are reused for new projects. This is a natural next step from coal plant closure. If we, Navajo, do not take advantage of these new projects, someone else will. There's a window of opportunity in which you have these idle transmission lines. If Navajo people don't make use of the transmission lines, somebody else will because they're very valuable. Transmission lines are a part of the infrastructure that will not be decommissioned because of how valuable they are. I ended up acquiring this as my next focus because of the closure of the plant.

So far, I'm working on reestablishing relationships between the Navajo Nation and the utilities. For instance, the city of LA still has ownership share to about 470 megawatts of transmission that's coming off of the NGS plant. I wanted to continue to use this transmission now available from the NGS closure and commit it to renewable energy right away. One, so the Navajo Nation continues to be a partner with this utility, and two, so that nothing but renewable energy would be put on those lines. In April 2019, I wrote a letter to the LA city council and pitched the idea of a Navajo-LA project. The LA DWP, the utility there that describes themselves as the largest utility in the nation, got approval from the city council to move forward and explore a potential partnership with the Navajo Nation.

The project would be situated on Navajo lands. The transmission lines would be used to move renewable energy to LA ratepayers. As we move forward with these new partnerships, we have to emphasize that these new agreements will be better than the previous agreements with utilities. Those agreements between NGS owners and Navajo Nation were really one-sided. Navajo practically subsidized these projects by selling for cheap or nothing at all. Going forward, these new agreements would include mutual benefit to the parties involved. The goal is to produce clean energy here on Navajo and put that clean energy on the transmission lines. The Navajo Nation and the city of LA would each benefit. There would be minimal impacts to the environment. And, most importantly, we wouldn't compromise the water.

I'm happy to say that this project is going forward. There were some delays with the pandemic, but the city of LA and Navajo Nation picked it back up in early fall of 2020 and they're moving along.

Once projects get going like this, it's important for groups to stay with it even though you feel like you're not being heard. I guess I'm kind of giving myself a pep talk right now because there's times when we go through long periods where I don't hear anything from President Nez's office and I don't hear anything from the LA city council. I understand when they're negotiating it's confidential. But I keep pushing information that I feel is important to make sure that this transition happens in a just and equitable way. I've been sending out these community benefits documents that we've created using the research of other people who have done community benefits agreements. We're telling the Navajo Nation to ensure that whoever the host is for this project, that they have some direct benefits, and that the Navajo Nation receives annual benefits from this project.

The host communities, whether it be the chapter or some other organization in the community, should have the opportunity to invest in the project, either to be part owners or have some kind of investment or share. In the past, anytime projects were proposed on the Navajo Nation, they would either buy the permits off of grazing permit holders and never give them back or they somehow put pressure on these people to give up their grazing permits. This is something that I feel is unnecessary. These permit holders should be allowed to defer their grazing for the duration of the project and be allowed to get some kind of compensation and annual payment for doing that. Why wrestle permits off of sheepherders? Why not just tell them to defer grazing for the duration of the project and then give the permit back to them when the project's done? This, I think, is better than outright taking it from them. These are the things that we're trying to make sure cross the finish line.

Another project is the reclamation of Black Mesa. It's so important to make sure that those lands are fully reclaimed. When I'm talking to congressional people or staffers in the Office of Surface Mining or the Energy and Natural Resources Committee, I let them know that if you can reclaim the lands to the highest standard so the people are able to use it again—for residential purposes, agricultural purpose, ranching purposes—that's transition right there. That's something that we're pushing really hard right now. We've been collecting documents from the Office of Surface Mining, and we're now at the point where we're asking for a congressional hearing on this very matter, because people who are in charge are not doing their jobs. And by this I mean the EPA [Environmental Protection Agency], OSM [Office of Surface Mining], and even Peabody. To do a poor job and leave the lands in a state where people can't live there anymore and can't develop it in any way except maybe for industrial use, what good is that? That's not a just transition.

By paying attention to and seeing reclamation through, we're going to help create jobs for mine workers who have been laid off and we'll be able to return the land so that the people can use it again. That to me is important, and that's a just transition as well.

Nicole Horseherder, Diné, is from the Black Mesa region of the Navajo Nation. She is a founding member of Tó Nizhóní Ání and has been an active member since its establishment. A graduate of the University of Arizona with a bachelor's degree in Family and Consumer Resources, she received an MA in Linguistics from the University of British Columbia, Vancouver BC Canada. Outside of Tó Nizhóní Ání, Nicole enjoys her time with family, horses, ceremonies, and traveling.

BUILDING A REGENERATIVE FUTURE

COOPERATIVES, INDIGENOUS FOODWAYS, AND LONG-TERM THINKING

BY LILIAN HILL

As told to Alastair Lee Bitsóí and Brooke Larsen
Novemeber 2020

The work that I'm currently involved in stemmed from my upbringing here in my own community, the Hopi community. I grew up close to the land, but also in a situation where my family didn't have a lot of resources living on the reservation. My parents were both hunters and farmers and lived a very simple life. We didn't have running water or electricity, and we didn't travel that much outside of the reservation. I had to learn how to be self-sufficient and how to participate in the ceremonies and the culture that my people have cultivated over many, many generations. Growing up in the village with community, with extended family, with clans, prepared me for the work that I'm doing today, which involves navigating the complex scenarios that we find ourselves in as a tribal community and as a humanity.

I grew up around my great-grandparents and others who built their own homes, who grew their own food, who established relations and trading systems with other tribes to access food and resources locally and regionally. The generations who came before me set a lot of foundational elements for rebuilding community in a sustainable and generative way.

Of course, that has all been interrupted by colonization and by US imperialism and the imposition of foreign Western values systems, which contributed to the dysfunction that our communities are facing today—not only tribal communities, but the global community. As a young Indigenous person who is trying to navigate these complex situations, I always go back to the teachings and the traditions that have been passed on to me. I try to remember what those are and try to honor those ways as I move forward.

When I was around sixteen, I was exposed to some elders in my community that were imparting certain cultural practices on to my generation. They were challenging us to become leaders for the community, to learn the ways, be initiated into the societies, and accept the challenges that were coming our way. Hopi leaders acknowledged that humanity was on the path of destruction, of destroying the earth and destroying our own humanity. We see that with the climate catastrophe that we're in right now. The elders were trying to provide an understanding of where we are at this point in time and how the conditions that we're going to experience are going to take a lot of bravery. As young people, we needed to be able to strengthen ourselves spiritually, emotionally, and with the skill set that our ancestors imparted onto us to weather and endure. For myself, that meant digging deep, understanding my purpose, and not relying on these larger imposed systems. That led me on a journey to build my own home, to grow food, and to honor the teachings that my ancestors have left for me to live a self-sufficient life of dignity and purpose.

Many of our community members are living in situations of limited mobility and financial poverty. There was already a need within our communities before COVID-19 hit. When the pandemic emerged, there was little or no support for families without running water or electricity. And there was no support from the government to provide the necessary food and resources that would enable folks to shelter at home.

I was invited by the Navajo Hopi COVID Relief effort, and by others who were engaged in this work, to help develop the strategy to provide support to our communities. I developed safety protocols and volunteer training and set up systems where we could safely provide food and emergency supplies to villages. That work has been transferred to local nonprofits who have more resources to do that safely and effectively with paid staff.

I was involved with the Navajo Hopi COVID-19 Relief effort for the first five months. I saw that our communities are vulnerable and there are elements within our society that are fragile and dependent on outside forces and resources. It's great that mutual aid networks are trying to develop networks of support for others within our communities. There has been substantial investment into supporting our communities during this pandemic. But a lot of those resources are not being strategically employed to build infrastructure and systems that are going to endure beyond the pandemic. A lot of the solutions are short-term, emergency-based solutions. This network had, and still has, the opportunity to build infrastructure and collectives that could be long-lasting. But the ball is being dropped a little bit. And that's my own critique. Internal issues happen within that work that nobody talks about related to the fact that we are just so colonized. We sometimes drop the ball because of our own colonial thinking and ideas of how leadership should emerge and how resources should be utilized. It's important to rethink those structures and systems, but if we're not coming from that place, then we're just utilizing the colonized mentality and moving forward in that manner.

I'm providing guidance around how to develop systems that will endure beyond emergency relief supplies, looking at food cooperative models that could build our food system and build resilience within our community. That work has been important because much of the emergency response is basically raising tons of money and purchasing food and supplies from the global corporate food chain, which is not building resilience. It's not building any type of stability. The COVID relief effort at times looked a lot like government rations or government emergency relief. Are these strategies helping us or are they making us more dependent? The coronavirus revealed behaviors, patterns, and complex challenges that our community was facing pre-COVID and will continue to face post-COVID.

It's important for us to reflect on the issues that our communities face. What do we want to keep? What do we want to let go of? What are elements within our societies that we value? How do we want to see our communities grow? What do we want to see for our families and the next generations as we move forward? As a young person, as an Indigenous person, those are questions that I am continually debating and trying to bring up within my own family, within my clan, within our tribal community, and within the global community. We are complex living beings in a living world. Moving forward, building a society that is regenerative and not extractive is the challenge of our generation and the challenge that's going to come up in the next generations.

The work I've been engaged in is nurturing conditions that help individuals, communities, and societies develop solutions with longer-term thinking. Part of that has been developing ideas of food, housing, and building cooperatives. What's needed in order to shift our society? We've needed government and organizational structures. This work to build regenerative societies takes technical components sometimes. And it takes leadership.

I emerged into this work as someone who is very hands on, knowing different skills of planting seeds, saving seeds, growing food, building homes, raising children—all of these different practical skills that I was raised with and have cultivated over the years. More and more, I'm involved in administrative work and researching. We've been developing reciprocal loan programming, support for farmers, support for food cooperatives and local food economies, share programs where farmers and others can access tools and equipment to grow food. There's just so much work that has to be done.

Educating people about local food systems and cooperatives is important to do in community or in cohorts, providing space for folks to learn. Some of those different strategies are apprenticeship models, training models, or skills-sharing models where folks can be engaged in a hands-on manner. As humans, we're emotional beings. We are intelligent beings that are also sensory beings. We learn through putting our hands on seeds or putting our hands in the earth. Those are the experiences that are most meaningful to us, whether we can admit it or not.

For instance, this past weekend, a bunch of folks from my own community got together, and we harvested and processed chickens for food. That whole process of putting your hands on an animal, offering prayers and blessings to that animal, taking that animal's life, plucking the feathers off, and processing the animal itself reconnects us to life and our journey as human beings. Capitalism has taken away our connection to those basic foundational elements that make us human: processing meat, planting seeds, harvesting, taking care of corn as it's growing, or learning how to prepare food. Those types of experiences are necessary to strengthen ourselves spiritually, emotionally, physically, and mentally so we have the confidence that we can survive the challenges we're facing. If you can't grow your own food, if you don't know how to survive and be self-sufficient, then you don't see anything beyond your current reality, which is basically continually giving our lives and labor to systems that don't care about us and aren't reinvesting in our communities.

It's important for us to utilize the tools that we were left with, the tools that we currently have. In order to do that, we have to interface with others. At one point in time, my own community and other Indigenous communities were more conscious of the world around us and living as active agents within our societies and the natural world. That knowledge needs to be brought forward in building regenerative communities, because our farming systems, our social systems, are and continue to be generative in nature, meaning that they continue to proliferate and enforce this idea of life continuing. We need to remember those ways.

Our community has seen different cycles of drought where we were forced to migrate to continue our way as a people. We have been through those periods and we have different tools that our spiritual helpers have left us with. Those tools include seeds, spiritual ceremonies, and songs that we learned along our journey and continue to bring forth as we move forward through challenges.

Today, many of our communities are dependent on outside food systems, on transportation systems that we don't have any connection to or any type of decision-making process over. As subsistence farmers, as people who want to be self-sufficient and close to the land, we do see that these periods of drought are not only substantial, they're unpredictable. The predictions of scientists and the predictions of our own elders are telling us that we are going to be living in times of massive change and massive challenge. We will have to bring forth different solutions or call upon perhaps non-human relatives to help us endure. We're at that time, whether we want to admit it or not.

We are receiving less and less precipitation than we've ever seen within our lifetime or within weather records. We are facing extreme challenges in regards to climate change. We're not receiving any rain. Our seasons are just off, and we need to first realize that's the crisis that we're in. Our communities don't realize that, because there is so much abundance within the supply chain and we're not dependent anymore on the land itself to provide for us. We are suffering as a result of scarcity, as the result of climate change, and we won't see those stark contrasts until the supply chains are interrupted. Many of our traditional elders have already seen that, and they see that we're living in a drastic and serious time.

It's important to also realize that we are part of living systems that are resilient, diverse, and strong. I continually try to strengthen myself spiritually through expression of my own indigeneity, of my own songs, of recalling my own traditional stories, and within the seeds that I carry and steward. I ask for guidance from those spiritual forces that my ancestors left for me. I call upon my ancestors as I navigate this world. My ability and my confidence comes from the fact that despite all of the challenges that we face, I have control over my life. I know how to take care of seeds, I know how to plant the seeds, I know how to harvest my own food, I know how to hunt, I know how to build my own home. My people, my ancestors, have left us with the tools that we need to thrive despite the challenges that we face. Moving forward, those are the tools that will help us build our capacity and sustain us over the long-term. Also, our non-human relatives, the birds and the insects and the microbes that we don't see, the invisible forces, those are all around us and will strengthen us if we're able to recognize and acknowledge them.

A lot of the issues and problems that we face as humanity have come from a Western colonial mindset of control, domination, imperialism, and capitalism. As Indigenous people, as Hopi people, our perspective and our ideas are not included in that. Our Indigenous values and our culture are fundamentally at odds with this Western value system. That's why we are constantly protesting and putting our lives on the front line to protect our forests, to protect our water, to protect our land from this extractive mentality. As Indigenous people, it's important that we are engaged and involved in this larger conversation because this Western mainstream idea of capitalism needs to be challenged. We need to continue to reject colonialism, to oppose racism, and to bring our own thoughts and our cultural understanding forward.

Our Indigenous perspective may not always have the solutions to these complex problems. We need to be able to determine what is appropriate from our perspective while not becoming burdened with finding the solutions to these challenges that Western ideology has created. Sometimes we need to dig deep and understand what we can bring forth and how we can be in solidarity with other Indigenous nations, but we may not be able to be in solidarity with the forces that are working against us.

There's a tendency of this Western value system to say Indigenous people have the solutions to bring us out of this conundrum that we have created within ourselves. But ultimately, this mentality is an extractive mentality. And that has to do with power as well. As Indigenous people, we are brought forth to share solutions, but we are not directly engaged in the decision-making process. We are not empowered to share our perspectives because of the racism and the white supremacy that exists within the structures. If we don't have the power to make decisions that impact our communities and our societies, then we are just being tokenized, or our ideas and our perspectives are being mined to empower the dominant society.

We as Indigenous people need to work through the lens of an Indigenous approach. And to do that, we have to liberate ourselves or shed some of these things that have been imposed upon us. This innovation needs to come from us as a community, as individuals, digging deep and realizing our purpose. We also need to have the freedom to do that, the freedom to envision and work towards a just future. Part of that is the hard work of liberating ourselves. As an Indigenous person, I feel like I am burdened by all of these issues, by all of this baggage that has been imposed upon me and my people, and in order to see my own potential, I have to take some of those off. And in order to shed them, we have to recognize that there are burdens on us in the first place. That's the hard thing, to see that we can emerge into a new world.

We all have a role to play. It's important to share, to learn together, and to ask ourselves what we want to keep and what we want to let go of. We have to be engaged in conversations, find solutions, and adapt to challenges that we face. Understanding this dominant ideology of destruction and extraction, those forces that keep us from realizing our potential, is important to be self-governing, self-sufficient, self-determined, and sovereign.

Lilian Hill founded the Hopi Tutskwa Permaculture Institute to support Hopi youth and community in developing skills and practical experience to encourage a new generation of Indigenous earth stewards that will continue to carry on traditional agriculture traditions and cultural lifeways. Lilian and her husband, Jacobo, caretake a two-acre Permaculture Living Learning Site located in Kykotsmovi, Arizona. Both are certified permaculture designers, natural builders, high desert farmers, rainwater harvesters, and beekeepers. Together they strive to provide guidance and support the emergence and revitalization of Hopi foodways while building Indigenous food sovereignty.

A PRAYER IN THREE PARTS

BY ASHLEY FINLEY

"We don't want a seat at the table. Fuck the table. The table is full of oppressors. We want a blanket and a pillow down by the ocean. We want to rest."
—*The Nap Ministry*

"…come celebrate
with me that everyday
something has tried to kill me
and has failed."
—*Lucille Clifton*

The fruit bursts
Between my teeth
The juice trickles
From my lips to my chin
Dropping to my chest
Deep red
As if there is a bullet wound
But there is no pain

My mother,
80 years old and gentler
Than she was in her previous years
Looks at the moment on my chest
Touches her own scar in that same place
And asks if I am ok

I do not know how to answer her

But I say
"I am ok Mommy."
Anyway
And I want her to believe me.

I talk to Breonna and Toyin
Every single day.
We shared the same skin
The same sturdy bones
The same love for our mothers
All of us, named by our Ancestors
And made sacred in one way or another

In this realm we did not know each other
But as they have crossed into the next,
We are sisters

"I love you Sis. You deserved so much better. Rest sweet ones, I will carry on the fight"

And I know they hear me.

I do not have a daughter of my own, yet
But I must imagine her
Mahogany
Bursting with laughter
And love for the sweetness
Of ripe fruit

She will know how to find
The Big Dipper by tracing her fingers
On her own skin

She will smile at the Earth
And call it kin

Softness and ease will surround her
And she will pray as the sun rises and sets
To all those who came before
To the ones who named her and
To those whose songs
She carries in her heartbeat

And all of the pomegranates
Will burst for her.
Dripping down her chin
And off of her fingers
And it will never be blood.
And she will always be ok.

Ashe'.

Ashley Finley is a California native who is currently living in so-called Salt Lake City. She is a Black woman, a daughter, a sister, an auntie, a friend, a dog mom, a nature lover, a birth keeper, a medicine maker, a poet, an activist, and an educator. Her passion lies in the liberation of all colonized and oppressed people and in the facilitating of a return to sacred, ancestral knowledge that bloomed before colonization and capitalism.

HOLY

BY ASHLEY FINLEY

I am so tired.
But Also,
I am a well full of deep sighs
And unbridled love for a world
I am sure loves me back

I am barefoot and naked
And a deep hum of requiem
For the bones that existed
Before the ocean sank them

I am of the Earth
And of the sacred smoke too

My body is soft and lovely
And full of all of the good things
My grandmothers saved for me

I am an entire wet mouth
Swallowing the nectar of freedom

My hands stretch out
Snatching freedom
for my daughters

I am my spirit personified
I am the soul that was meant for me

The sunlight and the honey
Dripping from the moss

I am the reason for all of it.
All of it.
The whole fucking thing.

Biography on page 219.

TRIUMPH

CLIMBING OUT, STEPPING UP, AND RISING ABOVE

BY MELISSA-MALCOLM KING

As a queer, disabled person of color living in America, I know there is no commodity that can pay the price for injustice, inequality, and disparity. Long before the COVID-19 pandemic, I have been fighting to stay alive against all odds. My story is not unique, nor am I alone in this battle, nor my voice the only silenced one. My call is for all those who have had their voice stripped away by the world's social constructs. The familiar unheard screams resulting from the piercing bayonets of systemic racism provoke my desire to not only fight back, but also expose the harm caused by friendly fire from soldiers who claim to support the cause. These individuals stand in the shadows, failing time and time again to speak out against injustices or acknowledge the suffering of others, leaving me shattered with fragmented pieces of my soul.

As Dr. Martin Luther King Jr. said, "I must confess that over the past few years I have been gravely disappointed with the white moderate. I have almost reached the regrettable conclusion that the Negro's great stumbling block in his stride toward freedom is not the White Citizen's Counciler or the Ku Klux Klanner, but the white moderate, who is more devoted to 'order' than to justice; who prefers a negative peace which is the absence of tension to a positive peace which is the presence of justice."

For so long, I carried around the broken parts of me, taped together with the belief of being unworthy. I succumbed to the destiny of a life so far beneath my measure and heavenly call. Abuse, anguish, torment, and vile discomfort created personal destruction. Feelings of never being enough, which consumed much of my being, provoked inner turmoil and destroyed positive peace. I became like a lobster in a tank.

After lobsters become trapped for food, they are placed in tanks and must have their claws bound to prevent them from harming each other while trying to escape. One day, while purchasing items from a fish market, I observed several lobsters in a small surplus tank; one did not have its claws bound. This lobster climbed over and above the other lobsters to get to the top. However, once the lobster reached the top, the others moved around violently and forced the lobster back to the bottom. After a few tries, this lobster began to act like the others around them, bound, and stopped trying to fight for freedom.

This scenario is like an allegory for my life up until recently.

At each attempt to climb to the top and break the shackles of systemic racism, I found myself exhausted beyond measure and met with contention. Like the lobsters, I and my people were forced into a space of oppression. The fight against entrapment causes divisions such as colorism, internal hatred, and lateral violence. But the suffering and actions of other marginalized people is not what brings me down. While the ongoing COVID-19 pandemic has amplified these injustices, I've known for years, including among my ancestors, how systemic racism is the reason for the oppression. The blame lies squarely on white supremacy's shoulders for creating a culture that does not provide an escape or refuge from the generational effects of oppression.

Assimilation to Eurocentric culture and systems provided me with a false sense of community and connection to others who compressed my desire to succeed. My purpose became one of codependence. Low self-esteem and self-worth followed. While I did not start this war, I fought these societal battles with meager weapons and blocked micro-aggressions and racist actions with flimsy shields. I viewed myself as wearing protective armor in the battle, but in reality I just tossed stones and sticks against the weapon of mass destruction called white privilege.

Society assigned me a narrative that described a pathway to failure and left me thinking that freedom would never come. I became swallowed up in self-pity and acted out the narrative written for me from birth. Not only did I read the script, but I also became a method actor of life. I added my own tangled, twisted, and mangled adaptation to become more pleasing to the world around me. Mental slavery told me that the audacity to dream would result in the inability to survive. There came a time after years of captivity that I no longer had the desire to fight or even to live. I thought I was broken beyond repair. Climbing out of the tank required not only a change in thinking, but a change in action.

Like the lobster in the tank, I had become exhausted trying to climb to the top. Truth was, I was already free; it was my altered state of mind and societal brainwashing that prevented me from accepting my full potential and the most significant version of myself. Once I no longer succumbed to life in the tank, I began to realize that I am so much more.

The COVID-19 pandemic forced me, like so many others, to make changes in life for personal safety and survival. I learned that I had what I needed to survive, but I still had to make space and time to thrive. I began to live with a different thinking pattern thanks to a wonderful mental health counselor. After surviving lifelong abuse, I realized that neither my worthiness nor self-acceptance depend on or thrive in the places that society has attempted to trap me into or bar me from. I no longer desire to change myself to fit the frame, but instead have built a custom-made model. I live in a space of worthiness instead of despair. My voice amplifies now in honor of my elders and through the paths I create for others as a trailblazer.

I found that my various intersections of gender and sexual identity, race, disability, and multicultural communities did not blend together seamlessly, and in fact many voids and gaps appeared as parts of myself found acceptance in one area or the other but not in entirety. I have learned that self-love comes from embracing all parts of myself as equal and beautiful. As a result, I enjoy each of my chosen and birth families because each adds to the whole picture of my experience. Despite what societal constructs may dictate, I am not a part or pieces of anything; I was born whole.

Each of my diverse communities—queer, intersex, disabled, people of color—and the family members I associate with taught me that together our voices cannot be silenced any longer. Standing at an intersection in a road often requires deciding which side to take to move forward. My intersections do not require me to take sides but rather to endure and celebrate all that is me.

The greatest threat to humans today is not COVID-19, and likewise, there is no simple cure. Insidious hate begins with people living in a bubble of power with rose-colored glasses of ignorance and prejudice. Systemic racism intertwined with white privilege attempts to make a permanent financial, emotional, and social underclass for people of color. This results in the murder of my transgender siblings, and children ripped from parents at the border. The murder of George Floyd by police did not begin this process, but the public outcry exposed how white supremacy has been killing my people for generations. As the demand for equity has increased, it is my hope that humanity can finally begin creating a vaccination that injects awareness and a call to action.

I now demand that those in places of privilege move from throwing out life rings of equity and, instead, provide life jackets of equality. A life ring saves a drowning victim and brings them to shore again, whereas a life jacket prevents drowning from happening in the first place. Since equality is so far off on the shore, equity is the current remedy to saving the lives of those in marginalized communities.

To reach equality, I know we must first find security in equity. Glenn Harris, president of Race Forward and publisher of *ColorLines*, says, "Racial equity is about applying justice and a little bit of common sense to a system that's been out of balance. When a system is out of balance, people of color feel the impacts most acutely, but, to be clear, an imbalanced system makes all of us pay." Equity provides systems, intentional practices, and cultural narratives for equal opportunity so that those in vulnerable communities can find progression and prosperity. In doing so, my people and I can flourish in all our various intersections, increase social mobility, and find the greatest expression of our humanity.

Equity is not the happy ending. Instead, it is an opportunity to find healing and restoration with the motivation to obtain equality in the future. Equality provides safety, security, and, most importantly, proactivity in not only saving lives but also providing a quality life—something that is currently reserved for the privileged in the United States of America.

This requires those places of privilege to cast aside performative allyship and what Dr. Martin Luther King Jr. described as "a negative peace" preferred by liberal moderates. Stand up against hate in all forms even when it has not become a social media hashtag or requires that you stand in a place of discomfort. Allyship is the process of uplifting and saving lives at all times—not just when it is convenient or popular to do so.

I am not and never was the cause of this "international illness," but instead am a perpetual survivor of ships that have gone down and left me silently drowning, gasping for air. I have made it to shore, overcoming the riptides of inequality, but, despite my most extraordinary efforts, no amount of experience provides immunity, and the pain never goes away. I carefully hide the pain for survival. I watch as my siblings are drowning, and like Harriet Tubman, Sojourner Truth, Mary Prince, Sarah Mapps Douglass, and Ellen Craft, I am an abolitionist going beyond the shore, diving back in to save them.

On this exquisite journey, I have encountered those who claim that I fail to see the best in humanity, that my perceptions are not accurate or are delusions of grandeur. I carry the blood of my ancestors who came across as slaves, of my elders enchained in brutality disguised as white salvation and prosperity. I was born into a tank that became a prison, not realizing that I had my freedom papers all along.

I am more than my circumstances, more than what the world says I am or am not, a reminder that not only is life worth living, but it is mine for the taking. I am the sole owner of my dreams, the author of my destiny, and the scribe of my journey. I am not rewriting a narrative but reclaiming what was mine all along. I wish and desire that each of us can partake in this wondrous opportunity of reclamation. In doing so, I imagine a future without discrimination and hate. I imagine a time and place where each of us will recommit to erasing bias and no longer accept hate practices as the norm. I imagine a regenerative hope, with each of us sitting together united by a single purpose: eradicating injustice and uplifting the voices of those who have been silenced too long. In doing so, hope will never die, but live on as the actions of each generation improve upon and empower the next.

Hope flourishes when we allow our dreams to come through us. Many keep dreams within or alter them in fear of rejection. When we allow dreams to take life with a plan and purpose, our dreams come through us. Then we become trailblazers, leading the way for other dreams to become realities.

I see the hope of the future regenerating with each person living their most authentic selves, free of the barriers and chains of injustice. I see hope in the future with each voice not only making their dream a reality but connecting it to others' dreams to grow a cycle of healing and restoration.

Dreams are not simple notions of what could be in the unknown and mysterious future. The ability to dream allows me to cast aside the shackles placed upon me and build an authentic place of peace, prosperity, and genuine commitment to self. My dreams are the inspirations and aspirations that no one has a right or privilege to alter or deny. My dreams will not be dissolved or dismantled due to current circumstances or societal perceptions. Instead, I flourish despite the obstacles, barriers, and limits that the world gave me from birth. I am a conqueror, a warrior, and a freedom fighter. Now that I have taken my rightful place on the throne, I can command my destiny and determine my future.

Melissa-Malcolm King is a writer and advocate for the rights of queer people, disabled people, people of color, and survivors of domestic violence. Melissa-Malcolm runs Project B.E. S.A.F.E. and is a regular contributor to Exponent Magazine. *They serve on various boards, including the Disabled Rights Action Committee and West View Media.*

ABOLITION IS A WAY OUT OF ISOLATION

BY BRINLEY FROELICH

In 2011, I was halfway around the world in Japan teaching English at a preschool when Occupy Wall Street started gaining momentum. Even though I was living with my parents since my dad worked in Japan, I felt stuck in isolation hell. Traveling alone in a foreign country without knowing the language really made me a homebody, and I felt a deep yearning to be in the streets with my friends and chosen family. To help pass the time, I became obsessed with checking the news, feeling invigorated by hashtags and Twitter feeds. My loneliness, although it felt all-consuming at times, was abated by solidarity with the "99 percent."

When I returned to the United States, the movement had dwindled but the energy it left behind was deeply ingrained in me. Videos of police brutality never left my mind. When Trayvon Martin was killed and Black Lives Matter spread across the globe, I started to think more about what it means to be Black in this country.

As the years went on and police violence only seemed to get worse, I realized there was so much I didn't know and will never know. What I do know is that being white and financially stable protects me from a lot of the violence that exists in the world.

2011 was also the year I completed a two-hundred-hour yoga teacher training. In yoga, we talk about *ahimsa*, or nonviolence. When grappling with ahimsa, you first attend to the violence you commit against yourself. Then that expands to the violence you've committed against those close to you, which then expands to your community, and finally to the violence you commit against the world at large. While growing up as a Mormon girl from Utah, I was taught to categorize violence very differently.

I understand now how institutional violence has harmed my community. I understand that white supremacy and colonialism play a role in our military. I see how the military harms not only innocent civilians in other countries, but the soldiers, too, who thought they were doing the right thing by serving their country. I see a link between the idolization of the army and of the police, who serve as a kind of de facto domestic military. The more I see, the more work I have to attend to.

"How can we organize our communities in ways to make violence unthinkable?"

This question led me to learn more about the prison abolition movement.

People often say the police and prison system is broken and needs fixing. They talk about reform, as if changes within the system are the solution. Most of the people who fear a world without police are the people who are not already targeted by them. For many, especially for those who are structurally marginalized, a world with more police is a scarier thought than one without any. That fear is justified when you look at the history of policing in the US, whose roots are based in slave patrols from the South during the eighteenth and nineteenth centuries.

Black women such as Mariame Kaba and Angela Davis convinced me that the prison and policing system is not failing. The system itself is a source of violence, and it's working as designed. I don't see how we can solve violence with more violence.

The prison industrial complex intertwines with the government and other industries to "use surveillance and policing as a solution to economic, social, and political problems," as Critical Resistance, a prison abolitionist group founded by Davis and others, puts it. Abolition is a visionary movement that understands that our US society is rooted in white supremacy, the practice of enslavement, and exploitation of both land and labor. Reforms will never go far enough in a system that willfully chooses not to recognize or make amends for its past atrocities.

Critical Resistance envisions "the creation of genuinely healthy, stable communities that respond to harm without relying on imprisonment and punishment. [They] know that things like food, housing, and freedom are what create healthy, stable neighborhoods and communities."

After the national prison strike in 2016, I felt like I had been converted, and I started calling myself an abolitionist. The following year, I started to teach yoga in the women's facility at the Utah State Prison in Draper. I wanted to take my practice to a place that felt overlooked by my community and offer some sort of reprieve in a place where your body is no longer yours to move where it wills. I also wanted to touch part of the women's isolation and let them know that there are people in the community who are willing to show solidarity with their struggles.

After getting into prison abolition in 2016, I wrote a future manifesto for 2021 titled "WE HAVE PEACE! NO MORE POLICE!" to commemorate fifty years since the September 1971 uprising in the Attica, New York prison in response to the killing of George Jackson at San Quentin State Prison.

In my imagination, Departments of Peace and Liberation replaced Police Departments. Peace Protectors ensured that divestments were made from corporations and police, providing resources to invest in self-determined community programs.

The likelihood of a near future without police seemed impossible to me at the time, but it felt important to engage in the practice of imagining a better, more connected world.

Many people who identify as prison abolitionists share this understanding—that abolition won't happen overnight, or even possibly in our lifetimes. The work is seen as a "horizon," where we can envision generations ahead and start building the foundation of a new society based on interdependence and solidarity, rather than domination, exploitation, and isolation.

Teaching yoga and working on imagination projects felt like the start of something bigger, so in 2017 I cofounded Decarcerate Utah, a prison abolitionist collective that seeks to dismantle the prison-industrial complex and end the harm that it perpetuates. I started researching the current landscape of Utah's prison and policing system. Our state prison is relocating to a sensitive wildlife area, a disaster only exacerbated now with potential development of an inland port. Utah has the highest number of jail deaths per capita in the nation (at least 357 since 2000), and the second-highest incarceration growth rate. The deadliest year on record for police killings in Utah was 2018, according to reports from the American Civil Liberties Union.

Although I hoped for it, never did I expect an abolitionist movement to happen in Salt Lake City. Our work on prison abolition seemed on the fringe until the summer of 2020. Just a few months before then, COVID-19 exposed the holes in our system in new ways. People's isolation during the shutdown suddenly brought to surface anxieties and tensions that were brewing for years. Isolation can have different impacts on people, but what bubbled up for many was a shared sense of struggle.

Just a few days before the police choked George Floyd to death in Minneapolis, the Salt Lake City police had shot Bernardo Palacios-Carbajal in the back over thirty times. Suddenly staying at home was no longer an option. The streets began to flood with protestors, and it became quickly apparent that our community had had enough.

On June 2, I logged on to listen to the virtual Salt Lake City Council meeting and to leave a comment to defund the Salt Lake City Police Department. A $1 million funding proposal for public safety overtime and training was on the agenda in response to the protests against the death of Bernardo and in support of the larger uprisings against police violence across the nation.

I and my fellow organizers at Decarcerate Utah took up the call of other abolitionist groups and started a local campaign to defund the police and reinvest in communities. We created a Google doc with background information, scripts, and demands to aid and encourage like-minded people to call in to the city council meeting. I expected a few people would engage.

I started fist-pumping when two comments came from people who resonated with our message and shared their own version of it. The comments kept coming, almost all of them sharing a unified message from our script to defund to the tune of $30 million and put that money instead toward vastly underfunded social programs. After years of organizing, it seemed like suddenly people were not only listening to us but resonating with our message! I was taken aback by the response.

The meeting went until almost 1:00 a.m., with thousands of emails more to process. The final budget vote was pushed back another week, which I took as a cue that we had more time for city council members to engage with us and learn more about our demands and concerns.

As protests continued to happen almost daily in Salt Lake and across the country, I saw more and more handwritten and printed signs calling for the defunding of police. My hope for a better world happening within my lifetime shifted after seeing how much support the community gave to amplifying this message.

After the first public hearing, news reports indicated that the city council was seriously considering our calls to action. As we continued to encourage people to send comments, sign petitions, attend various events, and grow our audience, I heard more reports, including radio interviews with the mayor, city council members, and the chief of police who were wondering: Where did our demands come from?

Decarcerate Utah is relatively easy to find and contact. Plenty of people who organize there would happily answer that question, along with the other common concern: what "defunding the police" actually means. Instead of asking organizers and activists how the $30 million figure was determined, the common conclusion was that it had come "out of thin air." Later, in the fall, the slogan of "defund the police" would be blamed for causing political divisiveness. Some people even went so far as to say that the phrase meant something else. Yet everyone who participated that summer at protests seemed to be in agreement that defunding means taking money away from police department budgets—or divesting—and diverting that money into services in the community that are desperately under-resourced. Defunding the police is part of a divest/invest strategy propelled by Black radical activists to invest in holistic programs that can serve marginalized populations rather than criminalize them.

To clarify: A 30 percent cut in the Salt Lake City police budget was both an aggressive and reasonable demand and is a necessary first step in transitioning away from our reliance on policing social problems. Defunding the police would allow a large portion of our budget to go to social services that would support people and cover their basic needs. When people have what they need, like a home, healthy food, a well-resourced school, health care, and access to services like therapy, we see crime rates drop. When people do not have what they need, they are driven to find ways to meet those needs outside of the systems that have failed them. The best way to promote a safer and healthier city, then, is to find the gaps where people's needs are not being met, and to meet them.

With that, we have to confront the question if the police are meeting our community's needs. What services do they provide to make sure people are housed, fed, or given health care? When they receive the bulk of our budget, what does that say about the community we want to create? What does our budget say about where our true values lie?

Scholar and prison abolitionist Ruth Wilson Gilmore says abolition is a presence—not just an absence of policing and prisons. With abolition as our practice, we plant new seeds, and from them grow supportive and regenerative communities instead.

In order to reduce police violence, we must reduce police power; but simultaneously, we also need to promote life-affirming services that will ultimately provide people with the tools they need to conduct their own healthy, creative, and vibrant lives. What we understand as "crime" are often acts of survival done in reaction to where our social structures fail to meet our needs: poverty, hunger, houselessness, broken interpersonal relationships, underfunded schools, lack of access to higher education and a profit-driven health-care system create conditions that make it difficult to maintain stability. In short, being marginalized by social, economic, and political structures often requires acts of survival that are then targeted by the police. Meanwhile, people in positions of power are rarely held accountable for the violence they cause. Instead of confronting the root causes of what drives a person to cause harm, we're reacting to something after the fact, calling it a crime, and placing someone in isolation. This villainizes people and creates an "other" that gets pushed out of society and into jail or prison, instead of trying to realize how we can safely interact with each other on both a micro and macro level.

What would our communities look like without prisons, and without police? When people first hear about the idea of prison abolition, their first response is often fear. "But what would we do about ______________ (fill in the blank)? What about those types of people?" they ask. Misunderstandings of how prisons and the police operate often lead us to believe they are the bedrock of safety and order. Scary stories are blasted in the media to make us believe that certain types of people are to be feared. The system has become so normalized that we rarely use our imagination to envision alternatives. What might our world look like without prisons and policing?

Prison abolitionists use imagination as a dynamic tool to propose different ways of interacting with others in the world. When we rely on the prison system to keep functioning as is, we fail to use the full capacity of our imagination. There's still no telling whether prison abolition will happen in our lifetime. The abolitionists before the Civil War may never have thought they would see an end to chattel slavery in their lifetime, yet that didn't stop them from constantly being engaged in that work.

It helps to start by recognizing how the criminal justice system operates. Prisons take a one-size-fits-all approach, arguing that the same method will rehabilitate people who have done anything from consume a drug to taking a life, and places people out of sight and out of mind. We presume this isolation promotes public safety or even rehabilitation. Contradictory to this, it's apparent from the isolation of the pandemic that we are social creatures that need to work in communities in order to heal.

Isolation fails to address the needs not only of the victim of a crime, but of the family or friends of the person who is locked away. Before it even gets to that point, when you are put on trial, you are encouraged to remain silent or deny any wrongdoing to evade prosecution. Instead of asking how this person who created the harm can make it up to the person impacted, our approach to it is essentially just to ask, "Who did this, and how can we punish them?" Often survivors express dissatisfaction or re-traumatization by working in this system.

To shift out of this, we could determine what kind of obligations come out of abuse or exploitation. Instead of seeking vengeance and punishment when something wrong occurs, we could be grappling with questions about how we can repair what gets broken, and what or who needs to be involved in that. We could re-examine what keeps us safe and encourage growth to those things that create healthy communities and interpersonal harmony.

Utah without prisons, then, would look completely different from how we see it today. This vision looks like people being in community with each other and sharing resources, taking care of one another and making sure that everyone has what they need. These are some of the conditions that allow people to make fully informed decisions for the benefit of not just themselves but of their entire network and ecosystem.

Here's what I imagine. Reciprocity, over self-interest, is a primary motivator of actions. In general, our culture would shift from individualism to solidarity. This would be inclusive and celebratory of everybody's differences. It would be the kind of community that recognizes that everyone hurts, that everyone deserves healing, and it would be comfortable processing the ambiguity of different circumstances where harm is happening. It would shift power dynamics so that people cannot use their position in society to exploit others who have less or no power over them.

When I talk about providing people with all their needs with no strings attached, it's often met with skepticism about where the money would come from. But that isn't the issue—the problem is our lack of imagination and willpower to assert that we have the tools to keep ourselves safe, and that we can determine the outcome of our own lives.

Consider that the prison relocation in Utah has a budget now exceeding over $1 billion, and extra funding for the police during Operation Rio Grande, a three-year plan to perform a homeless sweep, cost around $67 million for the first year of its application. This doesn't even cover the day-to-day operations of the police, jails, courts, or prisons.

When it comes down to it, we have to determine how much money we want to keep investing into systems of violence. At what point does enough become enough to transition to a society that is less harmful, more equitable, and fully supportive?

We can imagine something different. We can imagine a world where quality health care, with preventive health care as a priority, would be accessible to everyone at no charge. Relationship skills classes could be taught in public schools, and therapy could be offered as preventative care. With drug use and sex work decriminalized, people who choose to use drugs or make a living would not have to suffer from social stigma on top of the effects of addiction, trauma, poverty, exclusion and police violence.

We could enjoy being in a place where everyone has a living wage. College, trade schools, and universities could be offered as a public good, with everyone encouraged to attend at any age. All debt could be eliminated. Food and shelter could be a right afforded to everyone.

Caretaking roles could be recognized economically as valuable to the community. Teachers and social workers could be paid significantly more, and domestic work could be recognized as the necessity it is in this society and paid appropriately. Caretaking work could have economic and cultural incentives to encourage people to join their ranks.

I imagine every block with a community garden to encourage this shift out of isolation. I imagine these gardens as centers of creative participation. Think of the games, concerts, unions, classes, showers, funerals, or weekly dinners that could take place in these gardens. During the winter, a yurt or simple structure could cover the dormant soil, while neighbors could plan their plots together for the next season. People uniting to struggle together is how we can make our way out of isolation.

What do you imagine?

Brinley Froelich is a writer, embroidery artist, yoga instructor, and community organizer. She is also one of the cofounders of Decarcerate Utah. You can find more about her work at booforever.com.

A LOVE STORY

BY ESTHER MEROÑO BARO

Author's Note: I use he/him and they/them pronouns interchangeably for Luca in this story as they are too young to define their own pronouns, and Lauren uses they/them pronouns exclusively. I like the website mypronouns.org if you want to learn more about life-affirming personal pronouns.

"How do we give our children what they need to know to survive, and what they need to want to survive?"
—adrienne maree brown, Octavia's Parables podcast

e need to do something about the bottle."

"I know, I had the same thought last night."

"But we need a plan, it's not just going to happen."

I pull out my phone and search "how to wean toddler off bottle" with the desperate conviction of a frazzled, sleep-deprived parent at their wit's end—which I am, *we* are. Lauren and I have spent three consecutive nights waking up to answer Luca's cries for another bottle, the restless nights fueling sleepy-eyed tantrums during the day for both parents and our two-year-old. I'm desperate to figure out how to get our kid to sleep through the night.

Our pediatrician had recommended weaning early. Babies (and their parents) sleep better when they can learn to self-soothe, she said, and becoming dependent on a bottle could get in the way of that. But our initial problem was getting him to go to sleep on his own in the first place. A nighttime bottle had conveniently solved that issue and mitigated my refusal to let him "cry it out." I was way more willing to spend hours of my evening lying in bed with Luca sticking his finger up my nose, waiting to welcome sleep, than to let my heart break listening to him scream for me until he gave in to exhaustion. Fortunately, Luca was more than happy to be left alone in bed chugging milk until they passed out, and once the pandemic hit, the extra time in the evenings to process and chill out became critical.

So here we are now, dealing with the consequences of a short-sighted strategy that continued for far too long because #2020.

With Lauren looking over my shoulder as we sit in bed, I scroll through an article in *Fatherly* that advocates for the "cold turkey" method, with an emphasis on talking through it before tossing the bottles out forever.

"Yeah we should definitely start talking about it, that feels good. But cold turkey?"

We move on to another article in *Today's Parent* written in first person by a mom who swaps the bottle for a sippy cup of water while on a trip, with the excuse that "they don't have bottles here." She gushes about how pleasantly surprised she was when her kid never asked for one again.

"I don't think Luca's going to get over it that easily."

Our toddler is still initiating earthquake drills after the 5.7 magnitude quake that shook us awake one morning last March; he's obsessed with the possibility of running into "scary spiders" around the house after seeing Lauren go into red-alert when they found a black widow in our living room; and while it's December as I write this, he just stopped asking if there's fire in his closet from shrapnel that fell on the roof above his room during Pioneer Day fireworks in July… A small sampling of the "monsters" that walked the earth this year—or at least the ones Luca was consciously aware of and verbally coherent enough to recognize and express as unusual disruptions (the pandemic's been too much of a slow burn to register as an "event").

"We need a good story."

Stories have been a critical part of helping me survive and adapt to change throughout my life. I was born in Spain, moving back and forth between there and the US a few times from ages three to twelve as my parents struggled through economic instability, mental health crises, and marital strife. Despite the xenophobia they experienced in Utah and the barriers our undocumented status created, the pull of the American Dream—a core component of the worldview they'd converted to as members of the Church of Jesus Christ of Latter-day Saints—was strong enough to bring us back. After 9/11, the borders were more tightly secured, forcing a commitment one way or another. This is, of course, an oversimplified story to help move mine along. Both of my parents are complex human beings with their own reasons for making the choices they made, and I don't speak for them.

As I became more and more aware of the dysfunctional world around me, and without the tools to process my experiences, I turned to stories. My parents are gifted storytellers and have infused our lives with both real and imagined narratives. While my dad prefers to share stories from his own life as a way of expressing himself, and more recently, as a way to give us a more concrete connection to our genealogy, my mom was the Queen of Imagination. In my early childhood, she would tell my siblings and I colloquial Spanish bedtime stories, host puppet shows for me and my friends, and make up stories on the spot, always coming up with a variety of voices for each character. When we moved to Utah, she took us to the library every week with a laundry basket we'd fill to the brim with books, often reaching our check-out limit.

Books, movies, TV, and music became my coping mechanisms—my escape, fuel for my daydreams, a stable and safe portal into other people's minds, new worlds, infinite possibilities, and a way for me to piece together who I was and wanted to be. I was willing to try any genre, but favored sci-fi, fantasy, and first-person coming-of-age stories. I especially appreciated series that allowed me to build a relationship with the characters and their world over time, and would often choose books based on how thick they were, practicing self-discipline as I got to the final chapters, trying to savor them slowly if it was an especially good book.

At the age of eight, I made up my mind that I wanted to be a writer, eager to express myself and be seen by other readers in the way I searched for and found parts of myself in the stories I read.

It wasn't until much later in life that I developed more conscious awareness of the power stories had to create meaning, weave entire belief systems, and inspire action—for better or worse.

Like most parents, 2020 had us attempting to make sense out of a continuous series of emotionally intense situations—a handful of emergencies, some of them life and death, and many of them dealing with our environment and health. Luca latches on to each explanation of "What happened?" with focused curiosity and an incredible long-term memory for bits and pieces of stories, sustained by repetitive call-and-response phrases that describe the most dramatic moments.

"Nana *fell*. Nana *fell*. Nana *fell*."

"Yes, Nana fell. And the firetruck came to help her."

"The *firetruck* came? The *firetruck* came?"

"Yes, the firetruck came, and she's okay now!"

"Nana's *okay*. Nana's *okay*. Nana's *okay*."

It's kind of poetic. The conversation doesn't move on until you've affirmed you hear, understand, and agree with him by repeating exactly what he says—which is often a variation of what you said. Basically, we spend a lot of time recalling and acting out scenes from life over and over again, and the bottles disappearing could very well be another paradigm-shifting event that plays out for months.

One such shift that required a reframe happened in early March, a week before pandemic lockdowns began in the United States. Luca had been struggling with a mild cold for weeks when he went into respiratory distress while visiting my mom in Midway, Utah, and was rushed down Parley's Canyon to Primary Children's Hospital by ambulance. We spent two nights in the PICU, Luca pumped full of sedatives, their chubby, twenty-month-old cheeks covered by a full-face mask attached to a ventilator. The doctors sent us home immediately after his oxygen levels went up and he could disconnect from all the tubes and wires, because the hospital was running at capacity. Since then, Luca's continued to have asthma-like symptoms that escalate into coughing fits and wheezing when the air is full of pollen or pollution. The air pollution in particular is proving to be a compounding public health crisis for residents across the Wasatch Front and beyond as human-driven climate change increases wildfires, and development projects like the Utah Inland Port threaten to further poison our sources of life.

After struggling to breathe for two days while strapped to a mask covering his entire face, Luca did not want another piece of plastic over his mouth and nose—even if it was just for thirty seconds. But a puff or two of preventative meds every day is how we manage the symptoms so we can play somewhat fancy free. And if there's anything that tears my soul more than letting my kid "cry it out," it's watching them struggle and scream as I physically constrain them. For the sake of both of us, we started calling the inhaler a "robot" and making silly mechanical noises and stiff dance moves when we pulled it out.

It took a couple of days, but our reframe was a hit! He started reminding *us* that it was time for robot. All the devices we've accumulated since then—humidifier, air purifier, nebulizer—have become part of the O_2 robot team. Of course, he doesn't like the inhaler robot anymore because it signals time for "night night" in the evening, and he's busy moving through Gene Sharp Junior's list of "198 Methods of No-Sleep Action" right now…

We thought we'd figured out how to manage the asthma-like symptoms when wildfire smoke blew across the region in late summer. Salt Lake City shot up the global Air Quality Index charts to place third among cities with the most toxic air. A small sniffle turned into another rush to the emergency room one night. This time, Luca spent a few hours excited to be hooked up to an even bigger robot while playing with his tablet when he should have been asleep in bed. As for Lauren and me, we were just as excited to leave without checking in to a room at the PICU. With a doctor's affirmation that we didn't overreact and a prescription for an at-home "super robot" (nebulizer) in our hands, we headed home as the hospital's morning shift began.

Outside, pieces of trash and dust were swirling violently against the workers making their way up the hill from the TRAX station. We drove past Liberty Park dodging branches, and as we got out of the car in our own driveway, the full-grown tree in my neighbor's front yard seemed to bend in half. Lauren and Luca stayed up to watch whole limbs from fifty-foot box elder trees crash down into our backyard. My bone-deep exhaustion turned the chaos into a foggy dream as I got back in bed listening to our patio furniture and who knows what else bump against the house. I swear I heard the Wicked Witch of the West shrieking as she rode the one hundred mph hurricane-force gusts of wind some later referred to as an "inland hurricane." Though, it's more likely it was Lauren trying to make the freak storm that killed a human and damaged homes across the city less scary for Luca.

When I woke up feeling heartbroken and defeated, the story I'd been carefully weaving for the past decade about the world and my role in it had completely unraveled.

I grew up being told to prepare for the world to end in my lifetime.

It sounds like a morbid existence when written out like that, but the particular story that accompanied my worldview was written in the early nineteenth century by a movement of Christian New Yorkers who imagined a loving and forgiving God—a cool dad who would send big brother Jesus to clean up our mess and bring a thousand years of peace before the final judgment day. The apocalypse was #goals: "If you're prepared materially and spiritually, you'll be celebrating when it comes," was the general vibe of my church leaders and fellow members.

My faith in this story began to fray when I was twelve and my parents finally divorced after our last back-and-forth move from Spain. We moved to Cedar City with my mom this time—a small, conservative town in Southern Utah where many people were kind and welcoming, but too many more were cliquish and fearful of "outsiders." I became anxiously aware of the way my family was treated, particularly my mom, who couldn't hide her thick Spanish accent and was forced to work two or three jobs at a time to make ends meet.

How could people who believed everyone was divine treat others like they were unworthy of dignity and respect? Well, it was right there in the scriptures. You fail to live by the book, you're going to be cursed with a hard life. It wasn't difficult to extrapolate from there that if people see you struggling, they're going to know you're a moral failure and treat you as such. And of course, no one had given me a comprehensive American history lesson or explained that racism, classism, misogyny, and xenophobia were still alive today. As I grew into adolescence, the shame grew with me, swelling every time I witnessed my family members get mistreated, or what felt even worse: every time our family became the neighborhood charity project.

But I knew something wasn't quite right. My parents had taught me that every human on the planet was a child of God, a sibling to Jesus, and had been "saved for these latter days" to be a beacon of truth, guiding others to accept the love of our big brother into their lives and step into their own divine role in fulfilling his plan to return to earth and bring peace. This paired nicely with the post-communist, "you're special and you can do anything you dream" message I heard from *Sesame Street* in the '90s.

Despite my dissolving faith in the Mormon Church, this was one part of the story I held on to.

After a series of experiences, which included waging war against the Cedar High PTA budget priorities as editor of the school paper; reading Gloria Anzaldúa's *Borderlands/La Frontera* in my Diversity in Literature class at the University of Utah; witnessing the Dreamer and Occupy Movements spark and grow during my college years; and then moving to New York City just as the #BlackLivesMatter movement went viral to support the Sanctuary Movement and other faith-based racial and economic justice organizers with their communications work, I began piecing together a new version of the story for myself and my role in it.

The arc of the universe bends towards justice, as Dr. King said; and what if this isn't the darkness of the tomb, but the darkness of the womb, as Valarie Kaur says; and if we organize the poor and work to repair the breach, as Rev. William Barber II says...well, we just might circumvent apocalypse and create heaven on earth, I concluded. And me? I'm going to be a beacon of truth, amplifying the words of the social justice prophets, guiding others to pray with their feet until we reach the promised land together.

I just hadn't taken into account that the land may not even be inhabitable by that point.

Our brainstorm starts in silence as Lauren and I consider the pathways and long-term consequences of the fantasies we'll weave to explain the disappearance of Luca's bottles. I think about the values and principles we could affirm in this process: letting go, change, adaptation, interdependence, reciprocity.

"Swap the bottle for a new friend?"

"To add to the pile? I don't know...he hasn't shown any interest in his stuffed animals."

We share ideas and eventually decide that we'll use Luca's upcoming change from a floor bed to a bunk bed as the catalyst—a change we're making as a potential safeguard against the inevitably bigger earthquake that's overdue on our fault line (this bunk bed can withstand two tons of weight). Before it arrives, we'll start talking about this new bed and how fun it will be to sleep closer to the moon and have a secret reading nook underneath to keep him cozy and safe. We'll tell him about the bed builders who are working so hard to make it just right and are asking us for one thing in return: now that Luca's got a big-kid bed, they need all his bottles so they can share them with babies who still need them.

Luca's his own human, so we can only guess at his response. While he's always concerned about the well-being of others and is often willing to share things that are meaningful to him (like the very last fruit snack), our main worry is that we may be creating unnecessary complications around the bed transition that will just compound the problem. We'll have to make the change irreversible and irresistible and, of course, create space for a grieving process around the bottles, if needed. To accompany and comfort Luca through the transition (and so we don't end up back in bed with him for hours every evening), we decide to get a stuffed animal of a character from a YouTube series he loves, which Lauren originally found to help him get over his fear of spiders.

June Jordan writes in *Revolutionary Mothering*: "Children are the ways that the world begins again and again," and adrienne maree brown boldly states in *Octavia's Parables* podcast, "Life or death depends on being able to slow down and have a different conversation—a conversation that brings you together."

While the spiritual alchemists and movement doulas among us are constant sources of wisdom and inspiration for me, it's through emotional, agenda-less conversations, wearing masks outside a campfire in my yard with friends, that I continue to make sense out of my experiences this past year. It's in organizing meetings over Zoom where we spend half our time just checking in that I develop strategy. It's in making the effort to be fully present with Luca while we play together, letting go of my sense of time and whatever "deliverables" I've been tasked with, that I've reconstructed my worldview and rebuilt the path to my divine purpose. Or at least, a purpose that is self-directed and accountable to my community.

I understand now that many of the changes our planet is going through because of the ways we have neglected and abused her are unstoppable. That *apocalypse* means "to reveal," and we've been going through it for some time. But I also continue to believe in heaven, because I've been there just as much as I've been through hell this year, in my friends' songs, my lover's touch, my family's unconditional support, an icy mountain lake, Luca's laugh.

My role is to make sure Luca knows how to keep his eyes wide, his heart open, and his feet moving. I'm going to tell them stories about the monsters they might face, along with the infinite possibilities that building beloved community creates for getting a good night's sleep; stories that celebrate the miracle of all life and the magic of self and community transformation; stories that spark curiosity, wonder, humility; stories to remind them that love is their birthright; stories to give them the desire to move towards life, one step at a time, whatever lies ahead.

I know that Luca will need a village of people, not just prophets, telling those stories from their own experiences and ancestral histories to become full-bodied narrative experiences. Hands to hold, lights to follow on his journey to realizing his own divine purpose—especially through the moments when he feels lost. Which is why my commitment to this place, this people, and my child is to be a part of building movements that can turn *everything* over for our babies to plant new worlds with their stories and dreams.

And that's a whole other piece of the narrative I have faith we'll weave together.

Esther Meroño Baro is a community organizer and multimedia artist based in Salt Lake City, Utah, who enjoys dreaming out loud with her friends, family, and community. She thinks bios are kind of silly, especially when written in third person. If you want to get to know me, email me at esthermerono@gmail.com.

THE DAY OF THE GREAT APOLOGY

BY FRANQUE BAINS

This story takes place in a make-believe world. Everything written is fictional except the fact that the theme of the last taped episode of Mister Rogers' Neighborhood *was forgiveness. May his message of love live on through us.*

"I love you, 143."
—Mr. Fred Rogers (1928–2003)

The day we watched Mr. Rogers say I'm sorry to his wife
after years of squandering their fortune
betting horses
became widely known as

The Day of the Great Apology

The video got leaked to Facebook and within minutes
viral
we couldn't stop watching

Maybe it was the sincerity of his eyes
or
the sad look of their puppy
cause the puppy knew that Mr. Rogers had done something really bad
our pets are always so empathetic you know

Whatever

All we knew
is that
Mr. Rogers
the kindest gentlest person on the planet
messed up real bad
and said I'm sorry

And everyone started apologizing immediately

The Next Day

I woke up
early the next morning
and drove to Roswell
to visit the gravesite of my matriarchs
and I wept
and doled out my apologies to my mother
I had judged her so harshly

And to my grandmother, and great-grandmother
for my audacity towards what I didn't understand

---

2 weeks later

Mr. and Mrs. Rogers sit down at the Red Table
and talk it all out

Jada gives them a heartfelt hug

Kanye came on next to apologize to Sway
Sway does know some of the answers

---

1 month later
My sister apologized
for stealing my favorite shirt
back in the 8th grade

She said she saw me cry
and that she loved me

That chick is always so late with everything

By that time 95% of the world had already doled out their acknowledgments and apologies
the sky was getting so blue
we didn't have to buy things anymore

---

6 months later

The Weeknd didn't get a chance to perform at the superbowl that year

Apparently professional sports leagues
are pretty messed up

So instead
they let him write a new global anthem
"A Promise to Love"

My favorite part...

Each day I'll do my best to love you
I'll be who I am
so you can love me

and you'll be who you are
so that I can see you

Makes me cry
every time

---

1 year later

Mr. Rogers holds a reunion
for the 20th anniversary
of the last episode of *Mister Rogers' Neighborhood*

The theme of that episode:
forgiveness,
of course

So many tear emojis on the instagram live feed
as Lady Elaine uses her boomerang
one last time

To change the world
for the better

Saying
Boomerang - Toomerang - Zoomerang
how sweet it is to use our magic
to help our friends

---

5 years later

Rituals have taken the place
of lies

and broken promises

Slow rituals that last for days
fill the sky with our tears and laughter

I've gained 5 pounds of pure joy

---

10 years later

I gave birth on the coldest day that year

We shoveled
so much snow
to make it to the birthing space

So cold
I couldn't bundle up enough
I don't think I cared
I just wanted to meet her

and speak the new promise
that we've learned to share
because of that great day

Each day I'll do my best to love you
I'll be who I am
so you can love me
and you'll be who you are
so that I can see you
and I know that sometimes
I'll hurt you
but I'll do my best
and slow down

and stop the harm
and listen
and say I'm sorry
so that I can love you
all the more

Biography on page 105.

ACKNOWLEDGMENTS

Thank you to the contributors for sharing your words and trusting us with your stories, imaginings, and strategies for building a new world. Thank you for your courageous and creative actions for justice and liberation.

Thank you to the land and water of the Southwest that continues to nourish us despite heatwaves, drought, and wildfires.

Thank you to the team at Torrey House Press—Anne, Kathleen, Kirsten, Maya, Michelle, and Rachel—for making this book possible.

Thank you to the People's Energy Movement, specifically Ashley Finley, Esther Meroño Baro, Franque Bains, and Hillary McDaniel, for creating an artist organizing project, Movement Building Medicine, to expand the impact of this book.

Thank you to the people who rose up throughout 2020 and inspired this book—from those who took to the streets shouting "Black Lives Matter" and "Defund the Police" to the health-care workers who faced the tragedy of COVID-19 daily. We hope in the new world that we're building, that you will always have the care, resources, and support you need to rest and thrive. We hope you don't have to be heroes, but can rather, as our contributors say, "just be."

Alastair's Thank Yous

Ahéhee nitsaago! Thank you so much for reading this anthology and supporting the amazing contributors, who remain on the frontlines in our communities across the Intermountain West. Their stories, work ethic, and love for our communities are why this anthology is so needed for the world to hear. They offer solutions for a New World Coming. Again, thank you contributors. You're lifelong friends, and may the Holy People and Creator continue to protect and heal you.

I also want to thank Torrey House Press and its staff and board of directors, and the other partner organizations, for making this anthology a reality. This includes Kirsten Allen, publisher for Torrey House, and her team of writers, editors, fundraisers, and organizers. Torrey House Press has been very important to my writer career, and I'm so grateful to be part of the family. Thank you.

Thank you to my co-editor, Brooke Larsen! She's an amazing writer and friend to work with on this anthology. There is no one else like her. She's smooth, collected, diligent, intelligent, and exactly knew where to put pieces in this anthology. She is the Sailor Moon for this anthology, a superhero. Thank you, Brooke, and I'm so relieved this is now published. Ahéhee for everything.

Most importantly, I want to thank the Holy People and Creator for offering this opportunity to breathe and live, which allows me to write and organize in the communities I come from. Whether that's in Dinétah (Navajo Nation) or any urban jungle my spirits leads, thank you so much!

Other thank you also goes toward our cultural healers, including those we lost over the course of this anthology. We lost many cultural healers, and without them I will not be where I am today. They're the knowledge holders that help keep me connected to the natural world. Ahéhee nitsaago for the collective knowledge I've gained through this role as co-editor for this anthology. You all helped me recover from COVID-19, and because of you I am here.

To shíma (mom) and shizhe'e (dad), thank you so much! You're everything to me and without your presence in this Glittering World, I'm not sure where I would be. Thank you for birthing and creating me, along with my brothers (Kelly, Thomas, Robinson, and Cameron) and sisters (Kabah, Erin, and Razhinder), and my nieces and nephews, who are all important to this work. Thank you for everything and loving us. Thank you to my extended family on both sides—my To'ahani and Kinyaa'aa'nii families. Ahéhee nitsaago. That's a wrap.

May you all Walk In Beauty—in front of you, behind you, on top of you, underneath you, and all the way around you. Hozho Nahaasdlii.

Brooke's Thank Yous

I wouldn't have been able to compile and co-edit this book without the relationships I formed and the lessons I learned over the past seven years through Uplift Climate. This community of young climate justice activists and organizers in the Southwest was my movement home for much of my twenties. Thank you to those who challenged me, nurtured me, shouted "yes" to radical visions for a better future, spent long hours at "retreats" working through strategies and theories of change, pushed me to more deeply examine my privilege and shift my roles, shared stories and laughs under star-lit skies, and always made magic happen in direct actions and camp kitchens. What a gift, that space to grow and dream alongside other young people with a shared love of the land, water, and people of the Southwest.

Thank you to our contributors who spent hours writing, editing, and sharing their story with us. Thank you for your vulnerability, creativity, and trust. It is an honor to work and create alongside you.

Thank you, Alastair, for persevering with me through the long process of compiling and editing this book. Thank you for saying yes to co-edit. Thank you for your humor, sass, passion, Leo energy, brilliance, and friendship.

Thank you to the team at Torrey House Press, who I've been fortunate to work with on various projects over the past five years. Thank you for nurturing me as a writer and editor. Thank you for filling a critical role by elevating voices from the Intermountain West.

Thank you to my partner, Andrew Butterfield, for your love, delicious cooking, master dish-washing (the revolution starts in the kitchen, as some say), and comforting embraces, especially on those long days when I often forgot to take care of myself. Thank you for the encouragement and care you give me every day.

I could write whole books with thank yous to my family. Thank you, Grandma, for your fiery spirit and unmatched silliness. Next to my desk hangs an old photo of you on strike as a public school teacher. Your fight for justice has always guided me. Thank you, Grandpa, for teaching me how to rebel and how to love the world. Thank you, Mom, for always encouraging me as a writer and endlessly supporting me every step of the way. Thank you, Dad, for teaching me to reach for the stars and fostering my love for the more than human world. Thank you to my sisters, Nicole, Gracie, and Davie, for your unconditional love. Thank you to extended and step family who helped raise me. So much love and gratitude to you all.

Finally, thank you to the Wasatch Mountains and high desert of the Colorado Plateau for being my home and my guide. Thank you to the water that still gives life against all odds. Some say the Colorado River is the hardest working river. In this new world, I hope you won't have to work so hard.

ABOUT THE EDITORS

ALASTAIR LEE BITSÓÍ

Alastair Lee Bitsóí (Diné) is a public health and environmental writer from the Navajo Nation. He is from the small community of Naschitti, which is nestled below the Chooshgai Mountains on the New Mexico-Arizona state line. Alastair is an award-winning news reporter for the *Navajo Times*, a 2021 Public Voices Fellow on the Climate Crisis, and currently the southern Utah reporter for the *Salt Lake Tribune*. He is a published contributor to two anthologies on Bears Ears with Torrey House Press, and he offers media and cultural sensitivity training for non-Native media outlets and environmental allies, who write about Indigenous-led conservation efforts and Indigenous peoples from the Bears Ears region. Alastair also crafts public health and environmental communication messaging under his consulting business, Near The Water Communications and Media Group. He has a master's of public health degree from New York University College of Global Public Health and is an alumnus of Gonzaga University.

BROOKE LARSEN

Brooke Larsen is a writer, community organizer, and narrative strategist. She calls Salt Lake City, Utah home, ancestral land of the Goshute, Shoshone, and Ute people. She is a recipient of the *High Country News* Bell Prize for emerging writers. Brooke has spent the past decade—most of her adult life—organizing with the climate justice movement. She cofounded Uplift, a youth-led organization for climate justice in the Southwest, and was a youth delegate to the UN Climate Change Conference in 2016 with SustainUS. As a descendent of Mormon settlers who colonized so-called Utah, Brooke focuses much of her organizing on wealth redistribution, truth telling, and white accountability. Story listening is also a central part of her work, including two long distance story listening tours by bike. Brooke has an MA in Environmental Humanities from the University of Utah and a BA in environmental policy from Colorado College.

ABOUT THE COVER ART

Mariella Mendoza is a multidisciplinary artist and media strategist with roots in the Andes and the Amazonian Rainforest. Mariella's work explores their experiences of queerness, migration, and displacement, while building a space for healing and resistance.

As a visual artist, Mariella's work has been featured in multiple local publications such as *SLUG Magazine*, *City Weekly*, and *Catalyst Magazine*.

You can find their writing online at *Everyday Feminism*, *The Body Is Not An Apology*, *Black Girl Dangerous*, and mariellamendoza.com.

PERMISSIONS

A version of "When You Displace a People From Their Roots" by Sunny Dooley was previously published in *Scientific American.*

A version of "A Reflection On Distance" by John Tveten was previously published in *Boatman's Quarterly*.

Sections of "Latinx Political Power: A Young Organizer's Journey" by Irene Franco Rubio were previously published in the *Mujerista*, *Prism*, *Teen Vogue*, and *ZORA*.

A version of "Filling a crucial need: Volunteers Serve on Frontlines During Pandemic" by Alastair Lee Bitsóí was previously published in the *Navajo Times*.

A version of "Abolition Is a Way Out of Isolation" by Brinley Froelich was previously published in *Catalyst Magazine*.

Torrey House Press

Voices for the Land

The economy is a wholly owned subsidiary of the environment, not the other way around.

—Senator Gaylord Nelson, founder of Earth Day

Torrey House Press publishes books at the intersection of the literary arts and environmental advocacy. THP authors explore the diversity of human experiences with the environment and engage community in conversations about landscape, literature, and the future of our ever-changing planet, inspiring action toward a more just world. We believe that lively, contemporary literature is at the cutting edge of social change. We seek to inform, expand, and reshape the dialogue on environmental justice and stewardship for the human and more-than-human world by elevating literary excellence from diverse voices.

Visit www.torreyhouse.org for reading group discussion guides, author interviews, and more.

As a 501(c)(3) nonprofit publisher, our work is made possible by generous donations from readers like you.

This book was made possible by generous gifts from David Folland & Betsy Folland, Patrick de Freitas & Lynn de Freitas, the People's Energy Movement, the Utah Sierra Club, the Sustainable Markets Foundation, and Patagonia. Torrey House Press is supported by Back of Beyond Books, the King's English Bookshop, Maria's Bookshop, the Jeffrey S. & Helen H. Cardon Foundation, the Sam & Diane Stewart Family Foundation, the Barker Foundation, Diana Allison, Klaus Bielefeldt, Laurie Hilyer, Shelby Tisdale, Kirtly Parker Jones, Robert Aagard & Camille Bailey Aagard, Kif Augustine Adams & Stirling Adams, Rose Chilcoat & Mark Franklin, Jerome Cooney & Laura Storjohann, Linc Cornell & Lois Cornell, Susan Cushman & Charlie Quimby, Betsy Gaines Quammen & David Quammen, the Utah Division of Arts & Museums, Utah Humanities, the National Endowment for the Humanities, the National Endowment for the Arts, and Salt Lake County Zoo, Arts & Parks. Our thanks to individual donors, members, and the Torrey House Press board of directors for their valued support.